AF779319

Application of Laser Processing Technology in Automobile Manufacturing

Application of Laser Processing Technology in Automobile Manufacturing

Guest Editors

Peilei Zhang

Mingwen Bai

Yingtao Tian

MDPI

Basel • Beijing • Wuhan • Barcelona • Belgrade • Novi Sad • Cluj • Manchester

Guest Editors
Peilei Zhang
School of Materials Engineering
Shanghai University of Engineering Science
Shanghai
China

Mingwen Bai
School of Mechanical Engineering
University of Leeds
Leeds
UK

Yingtao Tian
School of Engineering
Lancaster University
Lancaster
UK

Editorial Office
MDPI AG
Grosspeteranlage 5
4052 Basel, Switzerland

This is a reprint of the Special Issue, published open access by the journal *Coatings* (ISSN 2079-6412), freely accessible at: https://www.mdpi.com/journal/coatings/special_issues/Laser_Processing_Technology.

For citation purposes, cite each article independently as indicated on the article page online and as indicated below:

Lastname, A.A.; Lastname, B.B. Article Title. *Journal Name* **Year**, *Volume Number*, Page Range.

ISBN 978-3-7258-6758-5 (Hbk)
ISBN 978-3-7258-6759-2 (PDF)
https://doi.org/10.3390/books978-3-7258-6759-2

© 2026 by the authors. Articles in this reprint are Open Access and distributed under the Creative Commons Attribution (CC BY) license. The reprint as a whole is distributed by MDPI under the terms and conditions of the Creative Commons Attribution-NonCommercial-NoDerivs (CC BY-NC-ND) license (https://creativecommons.org/licenses/by-nc-nd/4.0/).

Contents

Preface

The global manufacturing sector is undergoing a significant transformation driven by technological advancements and evolving market demands, with the rapid shift towards electric vehicles (EVs) being a prime example. Laser-based material processing has become a cornerstone technology in this transition, providing the precision, efficiency, and versatility necessary for next-generation automotive production. This special issue focuses on the critical role of laser technologies in advancing automotive manufacturing, emphasizing sustainability, performance, and cost-effectiveness.

Laser processes such as welding, cutting, additive manufacturing, ablation, texturing, and marking enable the design and fabrication of lightweight, high-strength, and complex components for both EVs and conventional vehicles. Their ability to process a wide array of materials—metals, ceramics, polymers, and composites—establishes laser technology as a key enabler of innovation throughout the automotive value chain. The growth of EV manufacturing specifically highlights the demand for advanced joining techniques, efficient battery production methods, and lightweighting strategies, areas where laser systems offer significant advantages.

This collection features cutting-edge original research and comprehensive review articles covering theoretical advances and practical applications of laser processing in the automotive sector. Topics include laser welding of recycled materials and lightweight structures, novel applications in lithium-ion battery manufacturing, and recent developments in laser additive manufacturing and cutting. Lifecycle assessment and sustainability studies further provide valuable insights into the environmental impact and long-term benefits of laser-based production routes.

We sincerely thank the authors, reviewers, and editors for their contributions. We hope this volume serves as a valuable resource for researchers, engineers, and industry professionals in developing and implementing laser-based manufacturing solutions, inspiring continued innovation for smarter, greener, and more efficient automotive production.

Peilei Zhang, Mingwen Bai, and Yingtao Tian
Guest Editors

coatings

MDPI

Article

Mechanisms of Spatter Formation and Suppression in Aluminum Alloy via Hybrid Fiber–Semiconductor Laser System

Jingwen Chen [1,2], Di Wu [1,2,*], Xiaoting Li [3,*], Fangyi Yang [1,2], Peilei Zhang [1,2], Haichuan Shi [1,2] and Zhishui Yu [1,2]

1 School of Materials Engineering, Shanghai University of Engineering Science, Shanghai 201620, China; m355123147@sues.edu.cn (J.C.); m350121384@sues.edu.cn (F.Y.); peilei@sues.edu.cn (P.Z.); bingchengguxing@163.com (H.S.); yu_zhishui@163.com (Z.Y.)

2 Shanghai Collaborative Innovation Center of Laser of Manufacturing Technology, Shanghai 201620, China

3 Shenzhen Han's Lithium Battery Smart Equipment Co., Ltd., Shenzhen 518000, China

* Correspondence: wudi@sues.edu.cn (D.W.); lixt112247@hanslaser.com (X.L.)

Abstract: This study investigates the spatter suppression mechanism in aluminum alloy welding using a hybrid fiber–semiconductor laser system. By integrating high-speed photography and three-dimensional thermal-fluid coupling numerical simulations, the spatter formation process and its suppression mechanisms were systematically analyzed. The results indicate that spatter formation is primarily governed by surface tension and recoil pressure. In single fiber laser welding, concentrated laser energy induces a steep temperature gradient on the molten pool surface, triggering a strong Marangoni effect and subsequent spatter generation. In contrast, the hybrid laser system optimizes energy distribution, reducing the temperature gradient and weakening the Marangoni effect, thereby suppressing spatter. Additionally, the hybrid laser stabilizes molten pool flow through uniform recoil pressure distribution, further inhibiting spatter formation. Experimental results demonstrate that the hybrid fiber–semiconductor laser system significantly reduces spatter, improving welding quality and stability. This study provides theoretical and technical support for optimizing aluminum alloy laser welding.

Keywords: aluminum alloy; hybrid fiber–semiconductor laser; spatter; temperature gradient; molten pool dynamics; numerical simulation

1. Introduction

Aluminum alloys are widely utilized in the manufacturing of critical components in aerospace and new energy vehicles due to their high strength-to-weight ratio [1]. Laser welding has become a predominant method for joining aluminum alloys owing to its high energy density, rapid processing speed, and operational flexibility [2]. However, the inherent low viscosity and surface tension of aluminum alloys at elevated temperatures frequently cause process instability. Specifically, the evaporation of low-boiling-point alloying elements (e.g., Mg and Li) induces vigorous solid–liquid–gas phase transitions, increasing the propensity for welding defects such as porosity and spatter. Among these, spatter poses unique challenges due to its direct impact on weld morphology, surface quality, and downstream reliability. Despite decades of advancement in laser welding technology, significant challenges remain in the welding of aluminum alloys containing volatile elements.

Conventional single-mode Gaussian laser beams, with their concentrated energy distribution, frequently induce unstable keyhole dynamics, resulting in defects including spatter, incomplete fusion, and porosity. Of particular concern is the fact that excessive metal spatter not only compromises mechanical performance but also poses safety hazards. Moreover, compared with steel and other structural alloys, aluminum alloys are more susceptible to spatter due to their lower viscosity, higher thermal conductivity, and rapid melt pool response under laser irradiation. These physical characteristics amplify interfacial instabilities, making it more difficult to suppress spatter through empirical parameter adjustment alone. Therefore, a deeper physical understanding of spatter initiation and evolution is essential.

To mitigate these limitations, dual-beam hybrid laser systems have emerged as a viable solution [3]. Notably, the hybrid fiber–semiconductor laser system leverages unique energy distribution characteristics to address aluminum alloy welding challenges. Compared to single laser sources, this system—configured with coaxial or dual-beam fiber–diode arrangements—modifies laser energy distribution, attenuates molten pool temperature gradients, and enhances process stability, thereby improving weld quality.

Beginning in 2000, researchers initiated investigations into dual-beam hybrid laser welding for aluminum alloys to optimize process efficiency. Widespread industrial adoption commenced in 2011, following successful applications in copper alloy welding before being extended to aluminum alloys [4–6]. Glumann et al. [7] pioneered the implementation of dual 5 kW CO_2 lasers for hybrid welding, demonstrating stabilized process conditions and effective porosity suppression. Dual-beam systems enhance laser absorption and penetration depth, minimize porosity and spatter formation, and produce smoother weld surfaces with negligible porosity in full-penetration mode [8,9]. Furthermore, the strategic combination of wavelengths and spot diameters significantly affects workpiece absorption efficiency and weld morphology [10,11].

With advancements in fiber laser technology, Ishige et al. [12] conducted comparative studies of molten pool behaviors between semiconductor and fiber lasers, revealing that semiconductor lasers generate deeper, more stable molten pools. Maina et al. [13] employed a hybrid fiber–semiconductor laser system (1064 nm + 532 nm) for copper welding, with simulation results confirming increased penetration depth and processing efficiency. Zhu and Zhao [14,15] established that hybrid laser welding of aluminum alloys optimizes molten pool flow through coordinated heat conduction and convection, thereby enhancing process stability and weld quality. Zhao et al. [16] reported that coaxial hybrid wavelength laser beams (HW-HLB) exhibit superior power density compared to single fiber lasers (FLB), leading to plasma plume intensification and deeper keyhole formation.

Despite these advancements, current studies on hybrid laser welding remain largely concentrated on process optimization, with a notable lack of systematic research into the mechanisms of spatter suppression. In particular, limited attention has been paid to the interplay between thermal-fluid dynamics and spatter behavior in aluminum alloys, despite the fact that melt pool instabilities and interfacial force imbalances are key contributors to spatter formation. This study integrates high-fidelity multiphysics simulations with high-speed imaging to elucidate the influence of hybrid fiber–semiconductor laser systems on molten pool dynamics and spatter formation. Through parameter optimization, we aim to establish theoretical frameworks and technical guidelines for improving aluminum alloy welding quality, ultimately facilitating the intelligent development of laser welding technology.

2. Experiments and Methods

2.1. Hybrid Fiber–Semiconductor Laser Welding Experiments

The experimental investigation utilized 3003 aluminum–manganese alloy as the substrate material, with its detailed chemical composition presented in Table 1.

Table 1. The chemical composition of the 3003 Al-Mn alloy (wt %).

Element	Mn	Cu	Fe	Si	Zn	Al
3003	1.3	0.13	0.7	0.6	0.1	Bal.

This non-heat-treatable alloy is widely used in battery cases, vehicle bodies, and structural panels due to its excellent corrosion resistance, moderate strength, and good weldability. In particular, its relatively low melting point and high surface energy make it highly suitable for thermal processing using hybrid laser sources, as they promote stable melt pool formation under dual-beam irradiation.

The laser systems employed in this study consisted of two distinct laser sources manufactured by Han's Laser Technology Co., Ltd., Shenzhen, China: a single-mode fiber laser (Model: HL-WS-1000-G2) and a semiconductor laser (Model: HWD-2000), as shown in Figure 1a. These laser systems were integrated with a high-precision motion control platform, also produced by Han's Laser, which achieved exceptional positioning accuracy with a motion precision of $\leq$0.1 mm. This experimental setup provided precise control and reproducibility of the welding process parameters.

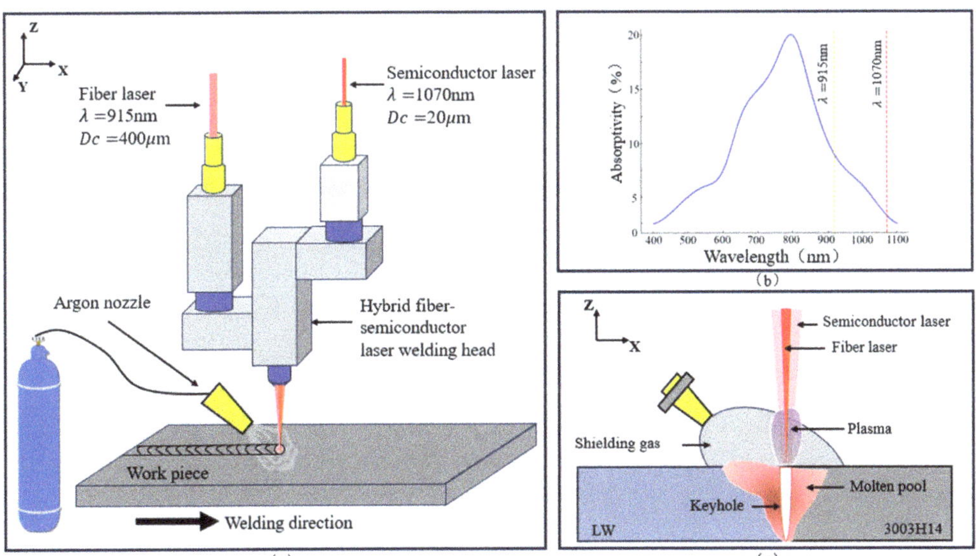

Figure 1. (**a**) Schematic of hybrid fiber–semiconductor laser welding system; (**b**) wavelength-dependent absorption characteristics of 3003 aluminum alloy [17]; (**c**) longitudinal cross-section of welding process.

Figure 1a shows the experimental setup for hybrid fiber–semiconductor laser welding. The integration of the two laser systems is achieved through a specialized pivoting dual-beam hybrid laser welding head (Model: HW-C-PT-1125001, Han's Laser Technology Co.,

Ltd., Shenzhen, China). In this configuration, both laser beams are collimated before being combined into a single coaxial output beam at the welding head. The optical system parameters are specified as follows: the semiconductor laser collimator has a focal length of 100 mm, while the fiber laser collimator has a 150 mm focal length, with both systems utilizing a common focusing lens having a 250 mm focal length. This optical arrangement produces a final output spot diameter of 0.3 mm for the fiber laser and 0.8 mm for the semiconductor laser. Figure 1b shows the wavelength-dependent absorption characteristics of 3003 aluminum alloy, while Figure 1c displays the longitudinal cross-sectional schematic of the welding process. The welding operation is performed under a high-purity argon (99.99%) shielding atmosphere, with a flow rate maintained at 20 L·min^{-1} to ensure optimal protection of the molten pool.

The welding parameters used in this study, including beam power, focal positioning, and scanning speed, were selected based on preliminary experiments designed to balance keyhole stability and spatter suppression under hybrid laser excitation. These parameter settings were consistently maintained throughout all trials to ensure experimental repeatability and comparability.

2.2. Energy Characteristics of the Hybrid Laser Beam

The optical architecture within the hybrid fiber–semiconductor laser welding head consists of two distinct subsystems, as shown in Figure 2a: the fiber laser beam transmission module and the semiconductor laser beam reflection module. In the semiconductor laser pathway, a 400 μm diameter beam is first collimated by lens A, then sequentially reflected by a mirror and beam splitter before reaching the final focusing lens. In contrast, the 20 μm single-mode fiber laser beam is collimated by lens B and transmitted through the beam splitter to the shared focusing lens. This configuration ensures both laser beams are coaxially aligned through the same optical element, achieving sub-micron focal alignment on the workpiece surface.

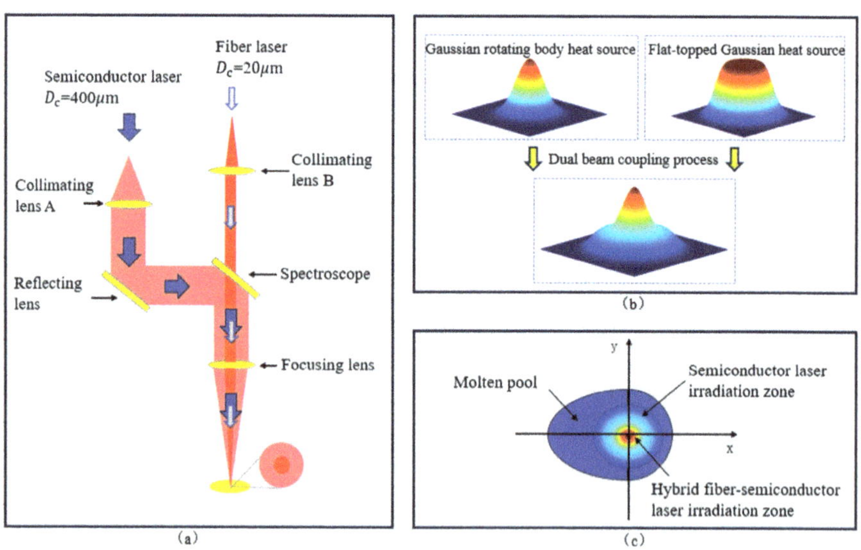

Figure 2. (**a**) Schematic of coaxial beam superposition in hybrid fiber–semiconductor laser system; (**b**) energy distribution profiles: single-mode fiber laser, semiconductor laser, and hybrid laser beams; (**c**) irradiation zone geometry of hybrid fiber–semiconductor laser.

Figure 2b shows the measured energy distribution profiles for three operational modes: the single fiber laser beam, the semiconductor laser beam, and the hybrid fiber–semiconductor laser beam. The irradiation characteristics are quantified in Figure 2c, demonstrating the spatially modulated laser energy distribution on the workpiece surface. The central high-intensity zone represents the combined fiber–semiconductor laser irradiation, while the concentric annular region corresponds to the peripheral 915 nm semiconductor laser irradiation. This engineered dual-zone energy distribution enables precise thermal management through controlled heat flux gradients and molten pool dynamics.

Based on the principle of coaxial laser beam superposition, a superposition model for multiple laser beams was established. Figure 3a shows the simulation results for the 1070 nm single-mode fiber laser beam, employing a Gaussian rotational heat source model. The key beam parameters—including divergence angle, Rayleigh range, and beam radius—were calculated using Equations (1)–(3):

$$\theta = \frac{\lambda M^2}{\pi\omega^2} \tag{1}$$

$$Z_R = \frac{\pi{\omega_0}^2}{\lambda M^2} \tag{2}$$

$$\omega(z) = \varpi_0\left[1 + \left(\frac{z}{Z_R}\right)^2\right] \tag{3}$$

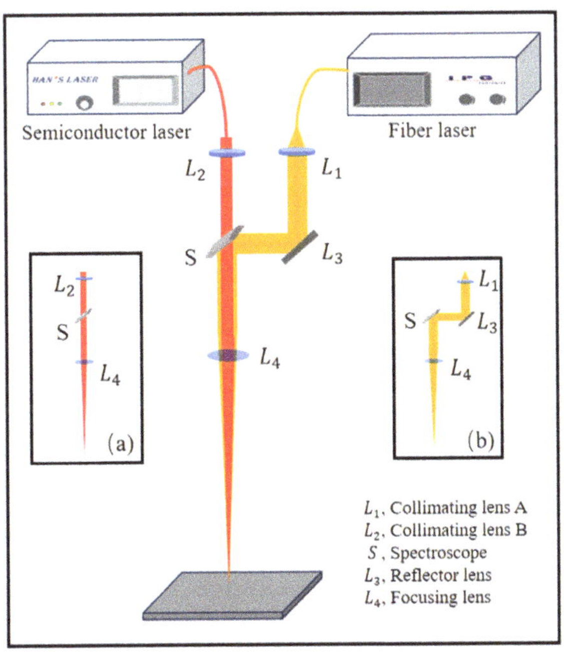

Figure 3. Simulation results of hybrid fiber–semiconductor laser beam: (**a**) 1070 nm single-mode fiber laser beam profile; (**b**) 915 nm semiconductor laser beam profile.

In the equations, θ represents the divergence angle of the fiber laser beam, λ denotes the wavelength, M^2 is the beam quality factor, ω_0 indicates the beam waist radius, Z_R corresponds to the Rayleigh range, $\omega(z)$ represents the beam radius at propagation distance z, and z is the axial distance from the beam waist.

Figure 3b shows the simulation results for the 915 nm semiconductor laser beam, employing a top-hat modified Gaussian heat source model. Building upon the standard three-dimensional Gaussian heat source formulation, we introduced a super-Gaussian coefficient N and divergence angle parameter θ. The super-Gaussian coefficient is calculated using Equation (4):

$$f(x) = \exp\left(-\frac{x^2}{\omega_0{}^2}\right)\sum_{k=0}^{N}\frac{1}{k!}\left(\frac{x^2}{\omega_0{}^2}\right)^k, N = 0, 1, 2\ldots \tag{4}$$

In Equation (4), N represents the super-Gaussian order coefficient, where k! denotes the factorial of the mode order k, used to control the beam profile sharpness in super-Gaussian formulations.

The simulation models for the fiber laser beam and semiconductor laser beam are coupled through non-sequential ray tracing. The focal length of the semiconductor laser beam was optically matched to the fiber laser beam's focal plane. Figure 3 presents the simulation results for the hybrid fiber–semiconductor laser beam system. Key parameter comparisons between the fiber laser and semiconductor laser are summarized in Table 2.

Table 2. Fiber laser and semiconductor laser excitation parameters comparison.

Parameter	Fiber Laser	Semiconductor Laser
Maximum power W	1000	2000
Wavelength λ/nm	1070	915
Fiber core diameter d_f/μm	20	400
Focal distance f/mm	200	200

2.3. *Multiphysics Modeling of Hybrid Laser Welding*

The hybrid fiber–semiconductor laser welding process involves complex multiphase interactions and material state transformations. To investigate the molten pool dynamics, we established a comprehensive three-dimensional thermal-fluid coupling model that explicitly accounts for the synergistic effects between the dual laser systems.

This numerical model incorporates three critical physical phenomena: gas–liquid–solid phase transitions, coupled heat and mass transfer mechanisms, and laser-matter interaction physics. The computational domain, illustrated in Figure 4a, encompasses both metallic and gaseous phases. Considering the symmetrical energy distribution characteristic of the hybrid laser system along the welding direction, we implemented half-symmetry boundary conditions to optimize computational efficiency.

The meshing strategy employs adaptive refinement to balance accuracy and computational cost. Near the laser interaction zone, we applied high-resolution hexahedral elements (0.02 mm × 0.02 mm × 0.02 mm), while peripheral regions utilize coarser grids (0.02 mm × 0.1 mm × 0.02 mm). Figure 4b presents this multi-scale mesh configuration, with Figure 4c providing a detailed view of the transition region between refinement levels. This approach maintains solution accuracy while significantly reducing computational requirements compared to uniform meshing.

In modeling fluid motion, the governing equations of mass, momentum, and energy conservation must be simultaneously satisfied [12]. The mass conservation (continuity) equation is expressed as

$$\frac{\partial \rho}{\partial t} + \nabla \cdot \left(\rho \vec{u}\right) = 0 \quad (5)$$

In the continuity equation, ρ represents the fluid density, and $\vec{u}$ denotes the velocity vector field.

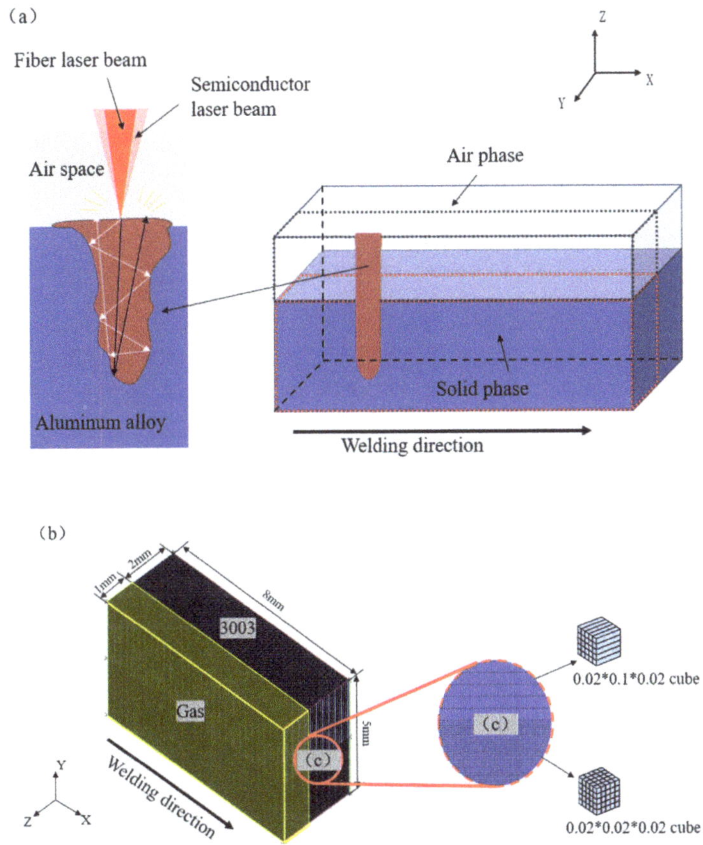

Figure 4. Thermo-fluid coupling model for 3003 aluminum–manganese alloy hybrid laser welding: (**a**) computational domain schematic; (**b**) mesh configuration results; (**c**) transition zone detail between refined and coarse mesh regions.

The momentum conservation equation is expressed as

$$\frac{\partial\left(\rho \vec{u}\right)}{\partial t} + \nabla \cdot \left(\rho \vec{u} \otimes \vec{u}\right) = \nabla \cdot \left(\mu \nabla \vec{u}\right) - \nabla P + \rho \vec{g} + \vec{F} \quad (6)$$

In the equation, μ is the dynamic viscosity coefficient, $\nabla \vec{u}$ is the velocity gradient, ∇P represents the influence of pressure variations in the fluid on its motion, $\vec{g}$ is the gravitational acceleration, and $\vec{F}$ is the external force term, representing external forces acting on the fluid (such as recoil pressure, thermal buoyancy, electromagnetic force, etc.).

The energy conservation equation is

$$\frac{\partial(\rho H)}{\partial t} + \nabla \cdot \left(\rho \vec{u} H\right) = \nabla \cdot \left(\frac{K}{C_P} \nabla H\right) + S \tag{7}$$

In the equation, H represents the enthalpy, K is the thermal conductivity, C_p is the specific heat capacity at constant pressure, and S is the energy source term, which corresponds to the laser heat source.

The VOF (volume of fluid) method is selected as the free surface tracking technique for welding pores. By solving the independent momentum equations and determining the volume fraction of each fluid within the control volume, the issue of tracking the gas–liquid interface during the transient melting process is effectively addressed.

The VOF equation is

$$\frac{\partial \alpha}{\partial t} + \nabla \cdot \left(\alpha \vec{u}\right) = 0 \tag{8}$$

In the equation, α represents the volume fraction, which indicates the volume ratio of a phase (such as liquid or gas) in a multiphase flow, where $\alpha \in [0, 1]$. In the computational model, the first phase is set as the 3003 manganese–aluminum alloy, and the second phase is set as the gas phase. When the liquid phase volume fraction of a cell is between $0 < \alpha < 1$, it indicates that the cell is in the gas–liquid mixing zone. Typically, cells with $\alpha > 0.5$ are considered to be in the liquid region, and cells with $0 \leq \alpha \leq 0.5$ are considered to be in the gas region. Therefore, in the entire computational domain, the fluid fraction of each grid cell will approach either 0 or 1, and the interface between the gas and liquid phases forms the gas–liquid free surface. Figure 5 illustrates this relationship.

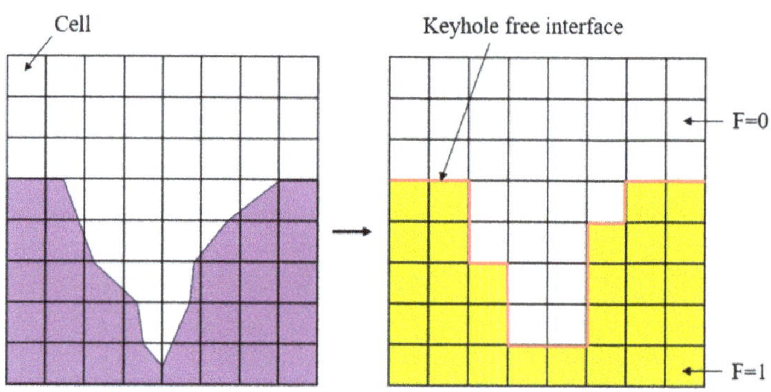

Figure 5. Definition of free surface in VOF.

At the free surface of the keyhole, the forces acting on it include the following: surface tension P_σ, recoil pressure P_r, Marangoni shear stress, static pressure, and other external forces. The keyhole wall maintains dynamic equilibrium under the combined effects of these forces [13]. The molten pool is governed by three dominant forces: thermal capillary convection, buoyancy force, and gravitational effects. The pressure balance in the normal direction of the free surface must satisfy the Young–Laplace condition:

$$P = P_\sigma + P_r \tag{9}$$

In the equation, P is the normal pressure at the free surface, P_σ is the surface tension, and P_r is the recoil pressure.

The surface tension is calculated using the following equation:

$$P_\sigma = \gamma \cdot \left(\frac{1}{R_1} + \frac{1}{R_2}\right) \quad (10)$$

In the equation, γ is the surface tension coefficient, and R_1 and R_2 are the principal radii of curvature of the free surface. The recoil pressure P_r is calculated using the following formula:

$$P_r = \frac{\rho v^2}{2} \quad (11)$$

In the equation, ρ is the fluid density, and v is the fluid velocity.

The surface tension coefficient of the 3003 aluminum–manganese alloy studied in this paper decreases as the temperature increases. The surface tension coefficient can be expressed as

$$\gamma(T) = \gamma_0[1 - \beta(T - T_0)] \quad (12)$$

In the equation, γ_0 is the surface tension coefficient at room temperature, β is the rate at which the surface tension coefficient changes with temperature, T is the current temperature, and T_0 is the room temperature.

At the keyhole wall, the vapor recoil pressure serves as the exclusive driving force for keyhole expansion and plays a pivotal role in pore formation during welding, substantially governing its morphological characteristics. Based on established research, the vapor recoil pressure acting on the keyhole during laser welding can be expressed as

$$P_r(T) \cong 0.54P_{sat}(T) = 0.54P_0\exp\left[\frac{L_v(T - T_b)}{\overline{R}TT_b}\right] \quad (13)$$

In the equation, $P_{sat}(T)$ is the saturation vapor pressure, P_0 is the atmospheric pressure, T_b is the boiling temperature, and $\overline{R}$ is the universal gas constant.

In the melt pool, the metal density varies with temperature. The buoyancy term is expressed using the Boussinesq approximation formula as

$$F_b = \rho g \beta_0 (T - T_l) \quad (14)$$

In the equation, g is the gravitational acceleration, β_0 is the thermal expansion coefficient, and T_l is the melting point of 3003 aluminum alloy.

Moreover, the choice of the heat source model will directly affect the consistency between the simulation results and the experimental results. According to previous work, due to the differences in energy density and power between fiber lasers and semiconductor lasers, there are significant differences in heat input, heat transfer modes, and energy distribution, which in turn affect the melt pool characteristics of the weld joint. The Gaussian distribution characteristics of the single-mode fiber laser lead to the formation of a narrow and deep keyhole in deep penetration welding modes. The temperature gradient on the surface of the melt pool is large, and under the influence of the Marangoni effect, natural convection of liquid molecules occurs, resulting in fluctuating flow and a tendency for keyhole collapse, thus forming a deep "V"-shaped weld seam. In contrast, the large-spot semiconductor laser primarily uses conduction welding, with energy distributed in a flat-top pattern. The liquid flow is more lateral, forming a bowl-shaped melt pool. In hybrid laser welding, the semiconductor laser promotes lateral flow in the melt pool, reduces the cooling rate, increases the overall temperature, and enhances the stability of the keyhole, ultimately forming an ideal Y-shaped weld seam and reducing the occurrence of spatter and porosity. Considering the energy characteristics of the two heat sources and the forming characteristics of the weld joint, a hybrid heat source model combining a Gaussian rotating body heat source and a flat-top Gaussian body heat source is established.

The described Gaussian rotating body heat source includes both surface and volume heat sources. The distribution function of the heat flux density of this heat source is described as follows:

$$Q_s(x,y) = \frac{\eta_1 q_s}{\pi {r_s}^2} \exp\left[-\frac{\alpha_r(x^2+y^2)}{{r_s}^2}\right] \tag{15}$$

$$Q_v(x,y,z) = \frac{9\eta_2 q_v}{\pi h_{laser} {r_s}^2 (1-e^{-3})} \exp\left[-\frac{9(x^2+y^2)}{{r_s}^2 \ln\left(\frac{h_{laser}}{z}\right)}\right] \tag{16}$$

In the equation, Q_s and Q_v represent the heat flux density of the Gaussian surface heat source and the volume heat source, respectively. η_1 and η_2 represent the efficiencies of the surface and volume heat sources, respectively, and q_s and q_v represent the laser power of the surface and volume heat sources, respectively. r_s is the radius of the heat source influence area, α_r is the parameter that controls the rate of heat flux decay, and h_{laser} is the depth to which the laser penetrates into the material.

The distribution function of the heat flux density for a conical Gaussian heat source is described as follows:

$$Q(x,y,z) = \frac{\eta q}{\pi[r_0(y)]^2} \exp\left[-\frac{x^2+z^2}{r_0(y)}\right] \tag{17}$$

$$r_0(y) = r_e + \frac{r_l + r_e}{y_l + y_e}(y - y_e) \tag{18}$$

In the equation, η is the efficiency coefficient of the conical Gaussian heat source, q is the laser heat source power, and $r_0(y)$ is the radius of the heat source influence area. The thermophysical properties of the 3003 aluminum–manganese alloy used in this study are listed in Table 3.

Table 3. Comparison between experimental and numerical simulation of molten pool size.

Physical Property	Symbol	Value
Density (solid)	ρ_s	2.73 g/cm^3
Density (liquid)	ρ_l	2.37 g/cm^3
Latent heat of fusion	ΔH_f	3.97×10^5 J/Kg
Latent heat of vaporization	ΔH_{vap}	1.04×10^7 J/Kg
Surface tension at room temperature	γ_0	0.87 N/m
Surface tension gradient	β	-1.6×10^{-3} N/m·K
Solidus temperature	T_s	916 K
Liquidus temperature	T_l	927 K
Boiling point	T_b	2743 K
Ambient temperature	T_0	300 K

The accuracy of this multiphysics model was confirmed through experimental validation. As shown in the Section 3, the simulated weld cross-sections demonstrated excellent agreement with experimental measurements, with all dimensional deviations controlled within 5%. Detailed validation data including weld width and penetration depth comparisons are provided in Figure 6 and Table 4.

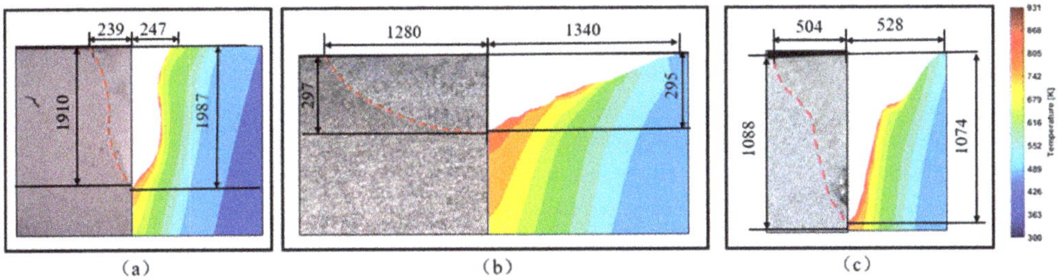

Figure 6. Comparative analysis of experimental and simulation results: (**a**) fiber laser power = 0.8 kW; (**b**) semiconductor laser power = 1.6 kW; (**c**) hybrid configuration: fiber laser = 0.6 kW + semiconductor laser = 0.8 kW.

Table 4. Comparison between experimental and numerical simulation of molten pool size.

		Weld Width	Weld Depth
Fiber laser power 0.8 KW	Experimental data (μm)	239	1910
	Simulation data (μm)	247	1987
	Errors (%)	+3.35	+4.03
Semiconductor laser power 1.6 KW	Experimental data (μm)	1280	297
	Simulation data (μm)	1340	295
	Errors (%)	+4.69	−0.67
Fiber laser power 0.6 KW, Semiconductor laser power 0.8 KW	Experimental data (μm)	504	1088
	Simulation data (μm)	528	1074
	Errors (%)	+4.76	−1.29

3. Results and Discussion

To systematically investigate the spatter suppression effect of the hybrid fiber–semiconductor laser system, the heat source models for single-mode fiber laser, semiconductor laser, and hybrid fiber–semiconductor laser were validated through weld cross-section comparisons. Three representative parameter sets were selected for validation: Figure 6a presents the comparison between experimental and simulated cross-sections for 800 W fiber laser welding at 70 mm/s with 0 mm defocus; Figure 6b shows the validation results for a 1600 W semiconductor laser; Figure 6c illustrates the hybrid laser welding case combining 600 W fiber laser and 800 W semiconductor laser. The simulation domain was selected at the position of maximum weld width. The results demonstrate excellent agreement between simulation and experiment, with weld penetration depth deviations within $\pm$5% (Table 4), confirming the accuracy of the proposed heat source model. This rigorous validation establishes a reliable foundation for subsequent investigation of spatter formation mechanisms and the development of effective suppression strategies in hybrid laser welding of aluminum alloys.

3.1. Analysis of Spatter Formation Mechanism

Under single-fiber laser welding conditions, spatter-induced pit defects frequently compromise weld bead surface quality, ultimately resulting in product rejection. Consequently, investigating spatter formation mechanisms becomes essential for developing effective suppression strategies. Figure 7 captures the spatter formation sequence through high-speed imaging. The process initiates when laser energy melts and vaporizes the metal surface, creating downward recoil pressure that establishes the keyhole (Figure 7a). Subsequently, ascending liquid metal droplets interact with cooler, turbulent molten pool fluid, constricting the keyhole aperture (Figure 7b). This interaction simultaneously reduces thermal energy at the keyhole apex. The inherent energy density gradient between the keyhole center and periphery establishes a temperature gradient that drives upward fluid transport along the keyhole walls through Marangoni convection. Protective gas shear forces deform the droplets while intensifying temperature gradients [18] impart higher ejection velocities (Figure 7c). At the keyhole exit, competing forces including inertia, surface tension, gravity, and recoil pressure collectively determine droplet trajectory (Figure 7d). The irregular molten mass eventually fragments under multidirectional stress (Figure 7e), with portions reintegrating into the weld pool while others solidify as discrete spatter particles (Figure 7f).

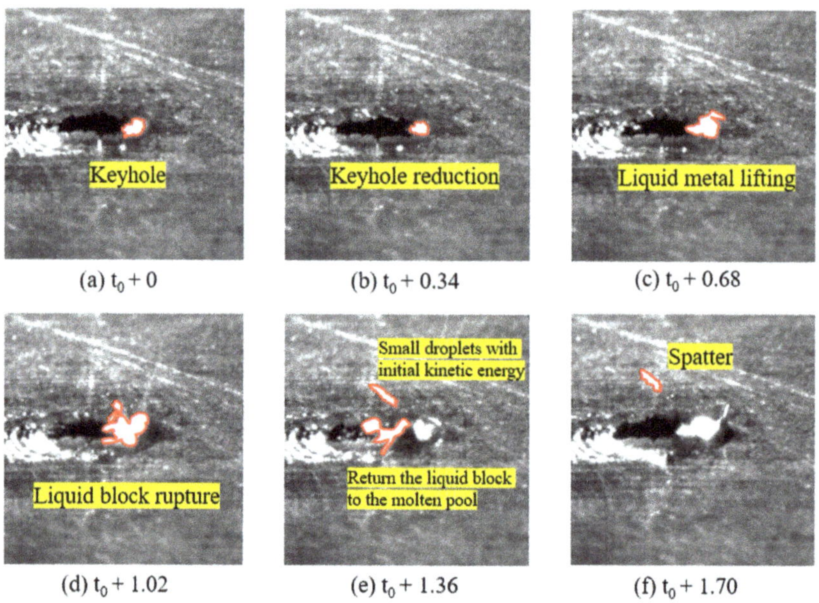

Figure 7. Formation process of spatter defect under high speed photography.

From the high-speed photography data in Figure 7, spatter formation is observed to initiate predominantly at the keyhole periphery. Thermocapillary instability may also contribute to spatter initiation by inducing wave-like deformation along the melt pool boundary under steep temperature gradients. These instabilities arise from non-uniform surface tension distribution and interact with vapor recoil and shear forces to disturb the molten interface. Related studies on aluminum alloy welding have reported similar unstable behaviors at the keyhole wall and melt pool surface under laser irradiation [19,20].

Figure 8 illustrates the vertical force balance acting on molten metal droplets near the keyhole, where the driving forces (recoil pressure and vapor shear stress) compete against restraining forces (gravity and surface tension). When the resultant upward forces exceed the downward restraints, droplet separation occurs: the upper portion ejects as spatter while surface tension draws the lower portion back into the weld pool.

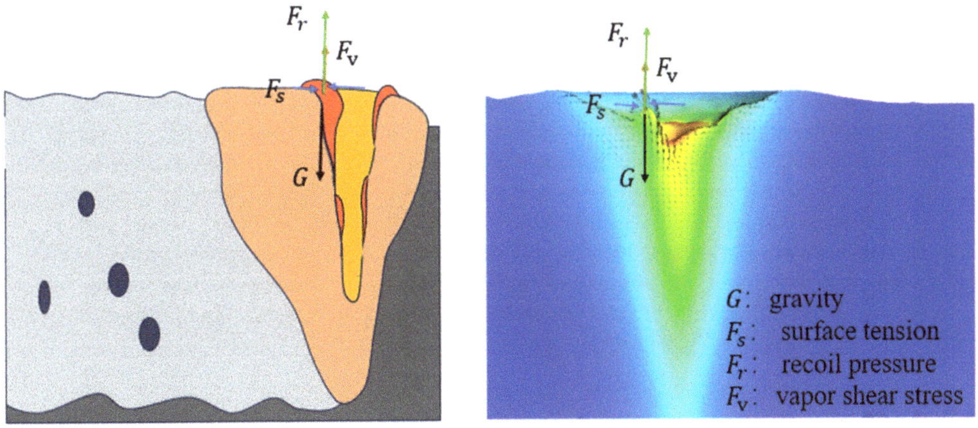

Figure 8. Force balance diagram of metal droplet in vertical direction around keyhole.

Figure 9 presents the formation process of spatter defects through numerical simulation. As shown in the figure, at the initial stage (t = t_0), laser irradiation rapidly establishes a high-temperature zone on the material surface, inducing melting and vaporization that forms a stable keyhole structure in the weld pool (Figure 9a). With continued laser interaction, the keyhole region temperature rises further. Metal vaporization within the keyhole generates downward recoil pressure, destabilizing the molten pool and driving upward fluid flow along the keyhole walls (Figure 9b,c).

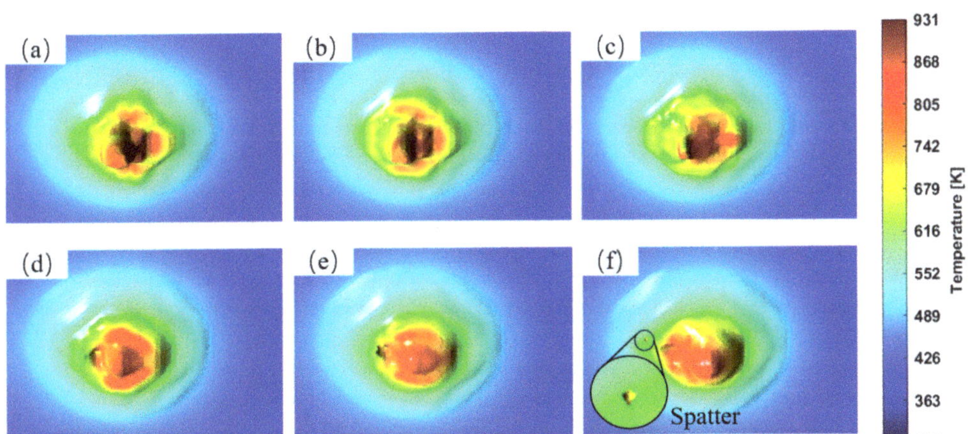

Figure 9. Formation process of splash defect under numerical simulation: (**a**) t = t_0; (**b**) t = t_0 + 0.34 ms; (**c**) t = t_0 + 0.68 ms; (**d**) t = t_0 + 1.02 ms; (**e**) t = t_0 + 1.36 ms; (**f**) t = t_0 + 1.70 ms.

At t = t_0 + 0.68 ms, sustained recoil pressure and vapor shear forces promote liquid metal accumulation at the keyhole apex, where upward motion is counteracted by gravity and surface tension (Figure 9c). When the driving forces overcome these restraints, partial liquid ejection occurs, resulting in droplet fragmentation and spatter formation (Figure 9d–f). These simulation results exhibit strong correlation with high-speed photographic observations, explicitly capturing the fluid dynamics under competing influences of recoil pressure, gravitational force, and surface tension.

Comparative analysis with experimental data reveals the complex force equilibrium governing molten metal behavior at the keyhole boundary. The numerical model confirms that spatter generation stems from time-dependent interactions between four principal forces: recoil pressure, vapor shear stress, surface tension, and gravity. Particularly, recoil pressure and Marangoni convection dominate the droplet ascent and rupture process, enabling liquid transfer from the keyhole interior to the workpiece surface where final spatter detachment occurs. The potential contribution of thermocapillary instability to interface fluctuation further supports this mechanism [20]. The consistency between simulation and experiment validates the model's capability to reproduce the essential physics of spatter defect formation.

3.2. *Study on the Spatter Suppression Mechanism*

Previous research has demonstrated that hybrid fiber–semiconductor laser welding exhibits significantly superior spatter suppression compared to conventional single-mode fiber laser welding. Spatter formation represents a prevalent defect in welding processes, particularly in high-energy-density laser welding applications. The occurrence of spatter adversely impacts both weld quality and subsequent manufacturing operations. Consequently, the implementation of hybrid laser welding technology offers an effective solution for enhancing process stability and final product quality.

3.2.1. Surface Tension Effect

Energy distribution uniformity and effective regulation of molten pool surface tension. Studies have shown that hybrid laser welding optimizes the laser beam's energy distribution, significantly reducing the temperature difference between the laser interaction zone and the molten pool surface. This thermal control substantially decreases the intensity of the Marangoni effect—the surface tension variation caused by temperature gradients in liquid metals that typically drives localized convective flows and spatter formation [21].

In single-mode fiber laser welding, the concentrated energy density creates substantial temperature differences between the center and edges of the molten pool, leading to strong surface tension variations that induce vigorous fluid flow and spatter generation. The hybrid laser system, combining semiconductor and fiber laser sources, achieves more uniform energy distribution during welding. This optimized thermal profile reduces surface temperature gradients, thereby suppressing Marangoni-driven surface flows.

The temperature gradient comparison between (a) fiber laser and (b) hybrid laser welding in Figure 10 clearly demonstrates this mechanism. The fiber laser welding shows steep thermal gradients with a distinct high-temperature core and cooler periphery, while the hybrid laser welding exhibits more uniform temperature distribution with reduced thermal gradients. This direct visualization confirms that the temperature homogenization in hybrid laser welding weakens Marangoni-induced surface flows, leading to effective spatter reduction.

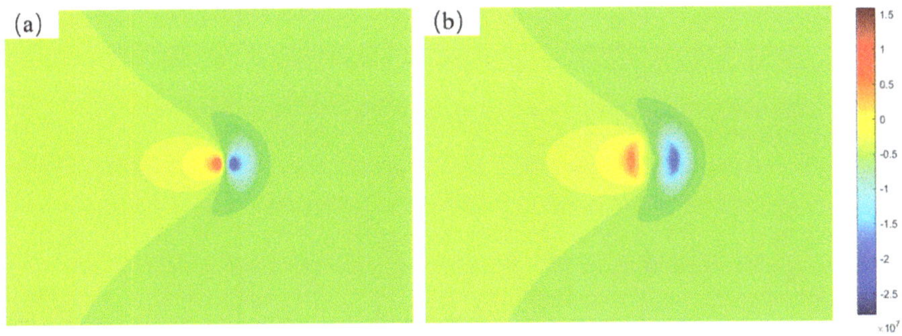

Figure 10. Comparative analysis of temperature gradients: (**a**) single-mode fiber laser welding profile; (**b**) hybrid fiber–semiconductor laser welding profile.

Experimental results demonstrate that hybrid laser welding significantly decreases molten pool surface flow velocities and reduces spatter generation compared to single-laser operations. This improvement stems from the hybrid system's ability to prevent localized high-temperature zones and subsequent surface tension imbalances. These findings provide fundamental theoretical support for enhancing welding quality while improving process stability in industrial laser welding applications.

As demonstrated in Figure 10, the hybrid laser system's ability to reduce temperature gradients serves as the first key factor for spatter suppression. This thermal control mechanism weakens Marangoni-driven convection, thereby decreasing the kinetic energy available for droplet ejection.

3.2.2. Recoil Pressure Effect

Beyond thermal gradient control, the hybrid laser system achieves spatter reduction through a second key factor: molten pool flow stabilization. Figure 11 directly compares the flow patterns under single laser and hybrid laser conditions, demonstrating how energy distribution uniformity prevents droplet detachment.

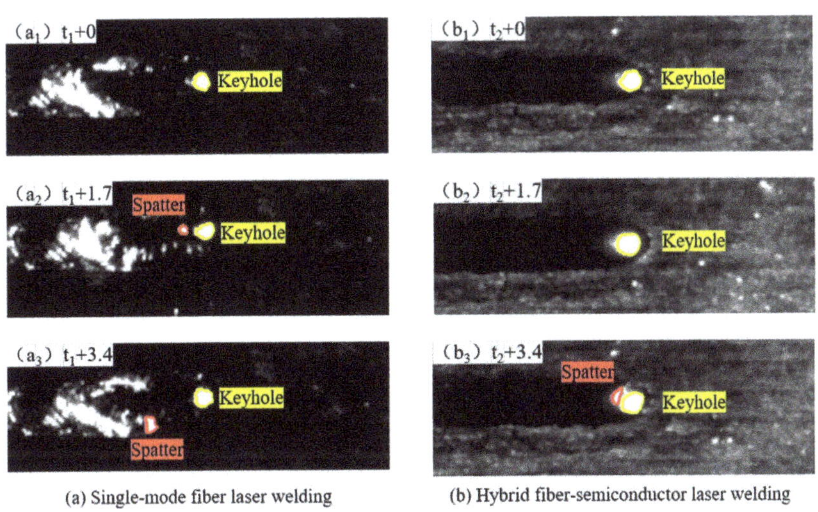

Figure 11. High-speed imaging comparison of molten pool: (**a**) single-mode fiber laser welding; (**b**) hybrid fiber–semiconductor laser welding.

In addition to surface tension optimization, recoil pressure plays a critical role in hybrid laser welding. This pressure originates from laser-material interactions, particularly the rapid vaporization induced by concentrated laser energy at the material surface. The resulting gas pressure increase drives molten metal flow within the weld pool.

Compared to single fiber laser welding, the hybrid system provides superior energy density distribution and broader power modulation capability. These advantages yield more uniform temperature and pressure profiles throughout the molten pool. The expanded pool volume and increased surface area help dissipate excessive recoil pressure, stabilizing metal flow while minimizing bubble formation and spatter generation.

The balanced energy distribution in hybrid laser welding enables gradual establishment of thermal and pressure equilibrium within the molten pool, preventing localized overheating and unstable flow patterns. This equilibrium enhances pool stability and reduces flow-induced spatter. Experimental evidence confirms that larger molten pool volumes promote smoother metal flow and decreased spatter formation.

The comparative high-speed imaging results in Figure 11 clearly demonstrate the process characteristics: (a) fiber laser welding exhibits pool instability with constricted keyhole dimensions, turbulent flow, and frequent spatter ejection, while (b) hybrid laser welding shows stable pool geometry with uniform keyhole expansion, laminar flow, and minimal spatter generation. These visual comparisons confirm that the hybrid laser's energy distribution achieves superior process stability through reduced thermal gradients and moderated surface tension effects.

Further refinements in power distribution and travel speed control can optimize gas pressure management, providing additional spatter reduction. These findings establish a theoretical framework for spatter control in hybrid laser welding and guide practical process optimization.

Spatter reduction results from two complementary mechanisms: temperature gradient minimization in Figure 10 suppresses Marangoni effects, while molten pool flow stabilization in Figure 11 inhibits droplet ejection. The hybrid system's energy optimization concurrently controls both aspects.

4. Conclusions

This study systematically investigates the spatter formation mechanisms and suppression strategies in hybrid fiber–semiconductor laser welding of aluminum alloys. Through the integration of high-speed imaging technology and a comprehensive three-dimensional thermal-fluid coupling numerical model, we have successfully captured and analyzed the spatter formation processes. The following conclusions were drawn from both numerical simulations and experimental observations:

- Spatter Formation Mechanism: High-speed imaging reveals that laser-induced metal vaporization forms a keyhole, where recoil pressure causes molten pool fluctuations. Droplet collisions reduce the keyhole volume, while temperature gradients trigger Marangoni convection, elevating droplets that subsequently rupture into spatter particles. Simulations confirm experimental observations of liquid metal dynamics under competing forces (recoil pressure, gravity, and surface tension), demonstrating that spatter generation results from complex multi-force interactions. These findings provide fundamental insights for welding process optimization.
- Surface Tension-Mediated Spatter Suppression: Hybrid laser welding effectively reduces spatter through optimized energy distribution and molten pool surface tension regulation. The process minimizes thermal gradients, thereby weakening Marangoni-driven flows. Experimental results demonstrate reduced molten pool flow velocities

and significant spatter suppression, ultimately enhancing welding quality and process stability. This approach combines semiconductor and fiber lasers to achieve superior energy distribution uniformity.

- Recoil Pressure-Mediated Spatter Suppression: The hybrid system suppresses spatter by controlling recoil pressure generated through laser-induced vaporization, promoting stable molten metal flow. Compared to single fiber lasers, the hybrid configuration provides enhanced energy density distribution and broader power modulation capability, enabling more uniform temperature and pressure profiles. High-speed imaging confirms improved process stability, showing well-developed molten pools with smooth flow characteristics and minimal spatter generation. Further optimization of laser parameters can enhance spatter control capabilities.
- Future Research Directions: Subsequent studies will integrate optical coherence tomography (OCT) for real-time spatter monitoring and employ machine learning techniques to analyze multi-sensor datasets for quantitative spatter prediction and classification. In addition, the effects of thermocapillary instability under evaporative recoil pressure on melt pool boundary fluctuations and spatter formation will be systematically investigated. This will provide deeper insight into the coupled fluid-thermal dynamics influencing spatter generation.

This study primarily focused on numerical modeling and qualitative analysis of dual-laser welding. Future experimental work should include quantitative measurements, such as melt pool dimensions and spatter counts, to further validate the simulation results. Such data would help establish clearer correlations between process parameters and weld quality. This integrated approach will advance understanding of spatter suppression mechanisms and facilitate welding parameter optimization.

Author Contributions: J.C.: investigating, writing—original draft. D.W.: conceptualization, methodology, reviewing, and funding acquisition. X.L.: data curation and modeling. F.Y.: software, validation. P.Z.: experimental testing. H.S.: supervision and editing. Z.Y.: writing—review and editing. All authors have read and agreed to the published version of the manuscript.

Funding: This research was funded by Foundation of Natural Science Foundation of China (grant number 52075317), Natural Science Foundation of Shanghai (grant number 24ZR1427100), and Shenzhen Major Science and Technology Projects (grant number KJZD20230923115121040).

Institutional Review Board Statement: Not applicable.

Informed Consent Statement: Not applicable.

Data Availability Statement: The raw/processed data and modeling codes required to reproduce these findings cannot be shared at this time as the data also forms part of an ongoing study.

Acknowledgments: The authors gratefully acknowledge all colleagues and collaborators who contributed to the completion of this work. We also thank the anonymous reviewers for their constructive feedback, which helped improve the quality of this manuscript.

Conflicts of Interest: Author Xiaoting Li was employed by Shenzhen Han's Lithium Battery Smart Equipment Co., Ltd. The remaining authors declare that the research was conducted in the absence of any commercial or financial relationships that could be construed as a potential conflict of interest.

References

1. Sadeghian, A.; Iqbal, N. A review on dissimilar laser welding of steel-copper, steel-aluminum, aluminum-copper, and steel-nickel for electric vehicle battery manufacturing. *Opt. Laser. Technol.* **2022**, *146*, 107595. [CrossRef]
2. Wu, D.; Zhang, P.; Yu, Z. Progress and perspectives of in-situ optical monitoring in laser beam welding: Sensing, characterization and modeling. *J. Manuf. Process* **2022**, *75*, 767–791. [CrossRef]

3. Zeng, D.; Wu, D.; Huang, H. Online identification of laser welding penetration through multi-photoelectric decomposition-reconstruction and shifted-windows-based transformer deep learning framework. *Measurement* **2025**, *247*, 116872. [CrossRef]
4. Hess, A.; Schuster, R.; Heider, A. Continuous wave laser welding of copper with combined beams at wavelengths of 1030 nm and of 515 nm. *Phys. Procedia* **2011**, *12*, 88–94. [CrossRef]
5. Helm, J.; Schulz, A.; Olowinsky, A. Laser welding of laser-structured copper connectors for battery applications and power electronics. *Weld. World* **2020**, *64*, 611–622. [CrossRef]
6. Zhao, Y.; Li, X.; Liu, Z. Stability enhancement of molten pool and keyhole for 2195 AlLi alloy using fiber-diode laser hybrid welding. *J. Manuf. Process.* **2023**, *85*, 724–741. [CrossRef]
7. Morawiec, M.; Różański, M.; Grajcar, A. Effect of dual beam laser welding on microstructure-property relationships of hot-rolled complex phase steel sheets. *Arch. Civ. Mech. Eng.* **2017**, *17*, 145–153. [CrossRef]
8. Punkari, A.; Weckman, D.C.; Kerr, H.W. Effects of magnesium content on dual beam Nd: YAG laser welding of Al–Mg alloys. *Sci. Technol. Weld. Join.* **2003**, *8*, 269–281. [CrossRef]
9. Sadeghian, A.; Nath, S.; Huang, Y. Quasi-continuous wave pulsed laser welding of copper lap joints using spatial beam oscillation. *Micromachines* **2022**, *13*, 2092. [CrossRef]
10. Bergmann, J.P.; Bielenin, M.; Feustel, T. Aluminum welding by combining a diode laser with a pulsed Nd: YAG laser. *Weld. World* **2015**, *59*, 307–315. [CrossRef]
11. Zhao, C.; Shi, B.; Chen, S. Laser melting modes in metal powder bed fusion additive manufacturing. *Rev. Mod. Phys.* **2022**, *94*, 045002. [CrossRef]
12. Faraji, A.H.; Goodarzi, M.; Seyedein, S.H. Effects of welding parameters on weld pool characteristics and shape in hybrid laser-TIG welding of AA6082 aluminum alloy: Numerical and experimental studies. *Weld. World* **2016**, *60*, 137–151. [CrossRef]
13. Fujio, S.; Sato, Y.; Takenaka, K. Welding of pure copper wires using a hybrid laser system with a blue diode laser and a single-mode fiber laser. *J. Laser. Appl.* **2021**, *33*, 042056. [CrossRef]
14. Rivera, J.S.; Gagné, M.O.; Tu, S. Quality classification model with machine learning for porosity prediction in laser welding aluminum alloys. *J. Laser. Appl.* **2023**, *35*, 022011. [CrossRef]
15. Jadidi, A.; Mi, Y.; Sikström, F. Beam offset detection in laser stake welding of Tee Joints using machine learning and spectrometer measurements. *Sensors* **2022**, *22*, 3881. [CrossRef]
16. Zhao, Y.; Zhan, X.; Ma, L. The energy synergistic mechanism of coaxial heterogeneous-wavelength hybrid laser beam and its effect on welding molten pool dynamics for aluminum alloy. *J. Mater. Process. Technol.* **2024**, *333*, 118605. [CrossRef]
17. Kaplan, A.F.H. Laser absorptivity on wavy molten metal surfaces: Categorization of different metals and wavelengths. *J. Laser Appl.* **2014**, *26*, 012007. [CrossRef]
18. Liu, G.; Zhang, Z.; Wang, H. Influence of laser welding defocus and penetration monitoring based on advanced optical sensors. *Optik* **2023**, *280*, 170811. [CrossRef]
19. Wang, L.; Zhao, Y.; Li, Y.; Zhan, X. Droplet transfer induced keyhole fluctuation and its influence regulation on porosity rate during hybrid laser arc welding of aluminum alloys. *Metals* **2021**, *11*, 1510. [CrossRef]
20. Ke, W.; Bu, X.; Oliveira, J.; Xu, W.; Wang, Z.; Zeng, Z. Modeling and numerical study of keyhole-induced porosity formation in laser beam oscillating welding of 5A06 aluminum alloy. *Opt. Laser Technol.* **2021**, *133*, 106540. [CrossRef]
21. Zhao, C.X.; Kwakernaak, C.; Pan, Y. The effect of oxygen on transitional Marangoni flow in laser spot welding. *Acta Mater.* **2010**, *58*, 6345–6357. [CrossRef]

Disclaimer/Publisher's Note: The statements, opinions and data contained in all publications are solely those of the individual author(s) and contributor(s) and not of MDPI and/or the editor(s). MDPI and/or the editor(s) disclaim responsibility for any injury to people or property resulting from any ideas, methods, instructions or products referred to in the content.

coatings

MDPI

Article

Investigation of the Inhibition Mechanism of Process Porosity in Laser-MIG Hybrid-Welded Joints for an Aluminum Alloy

Yucheng Xing [1,2], Feiyun Wang [1,*], Yong Zhao [1,*], Juan Fu [1], Zhenbang Sun [2] and Daxing Zhang [2]

1 School of Materials Science and Engineering, Jiangsu University of Science and Technology, Zhenjiang 212100, China; xyc8211@foxmail.com (Y.X.); fujuan@just.edu.cn (J.F.)

2 School of Materials Science and Engineering, Inner Mongolia University of Technology, Hohhot 010062, China

* Correspondence: wangfeiyunmse@126.com (F.W.); yongzhao418@just.edu.cn (Y.Z.); Tel.: +86-13815485895 (Y.Z.)

Abstract: In this paper, 4 mm thick 7075 aluminum alloy was utilized for conducting laser-MIG hybrid welding tests to investigate the correlation between the dynamic behavior of keyholes and process-induced porosity. Additionally, the generation and inhibition mechanisms of process porosity were elucidated. Utilizing a high-speed camera test system of our own design, the formation position and movement characteristics of keyholes in the molten pool under different welding parameters were captured using a "sandwich" method. The dynamic behavior of keyholes during the hybrid welding process was analyzed, and the porosity of each welded joint was quantified, revealing an intrinsic relationship between keyhole dynamics and aluminum alloy laser-MIG hybrid welding porosity. The findings indicate that variations in the defocusing amount can influence both the morphology and stability of keyholes in the molten pool, consequently impacting welding porosity. The dynamic behavior of keyholes under different defocusing amounts can be categorized into five types: no keyhole formation, collapse of the keyhole root, complete instability of the keyhole, instability of the keyhole root, and stability of the keyhole. At a defocus of +12 mm, stable keyholes were observed, and no defects in the welded joints were identified.

Keywords: 7075 aluminum alloy; laser-MIG hybrid welding; porosity; defocusing amount; keyhole morphology

1. Introduction

Aluminum alloys, with their low melting point, low density, easy processing, high specific strength, and excellent corrosion resistance, have become the preferred material for the development of lightweight products. They are widely used in automobiles, rail transit, aerospace, and other fields. laser-MIG hybrid welding technology combines laser and arc heat sources, offering advantages such as high welding speed, high efficiency, low heat input, and minimal workpiece deformation after welding. However, the complex welding parameters and high sensitivity to welding porosity in aluminum alloy laser-MIG hybrid welding limit its widespread application. In the 1980s, British scholar M. Teen first proposed laser-arc hybrid welding technology [1]. Since then, as an emerging material joining technology, it has been widely utilized in high-end manufacturing industries such as automobiles, aircraft, and rail transit due to its advantages of high welding adaptability and efficiency [2–4]. laser-MIG hybrid welding integrates laser welding and MIG welding, with the two heat sources superimposed and mutually influencing each other, resulting in a complex coupling effect. This approach not only leverages the advantages of both laser welding and MIG welding but also effectively compensates for their respective shortcomings. Reputable researchers all over the world have conducted extensive studies on laser-MIG hybrid welding technology since its inception. P. Leo et al. [5] investigated the impact of the proportions of laser and arc heat sources on the microstructural properties of laser-MIG-welded aluminum alloy joints. The findings indicate that the ratio of the laser heat source to the arc heat source significantly influences the macroscopic formation of welds, grain size, porosity, and magnesium burning loss in joints. In addition, it was found [6] that post-welding heat treatment can reduce the segregation of the element Mg, and the microhardness and tensile strength of the fusion zone can be increased when the heat treatment temperature is 350 °C. Aluminum alloys have physical properties such as high reflectivity and large linear expansion coefficients, which have a great impact on the formation of welds, the generation of welding defects, and the mechanical properties of joints [7,8].

Porosity is the predominant welding defect in laser-MIG hybrid welding of aluminum alloys and represents a critical challenge in the welding process. Its presence leads to a reduction in the effective cross-sectional area of welded joints, induces stress concentration, and has an impact on the mechanical properties and service performance of welded joints [9]. The formation of porosity is mainly affected by three factors: the solidification rate of liquid metal in the molten pool, the presence of hydrogen, and the stability of the keyhole. Domestic and foreign scholars have classified aluminum alloy welding porosity into "metallurgical" and "process" porosity [10]. The main reason for the formation of "metallurgical" porosity is that during the solidification process of hydrogen in the molten pool, its solubility decreases sharply with the decrease in temperature, forming pores with sizes of 50–200 μm, also known as hydrogen pores. Such pores are spherical, with small pore size and smooth inner walls, and can be eliminated by cleaning before welding and adjusting the protective atmosphere [11].

The main reason for "process" porosity is the instability of the keyhole during the welding process, also known as "keyhole" porosity. The shape of such pores is irregular, the apertures are large, and the inner walls are rough, which seriously affects the mechanical properties of the joint [12–14]. Huang et al. [15] conducted laser-MIG hybrid welding of 5083 aluminum alloy with the laser in front and the arc in front, respectively, and studied the influence of the sequence of laser and arc on the stability of the welding process and the mechanical properties of the joint. The results showed that the laser-arc hybrid welding (LAHW) mode had a more diffuse arc and a more stable keyhole, which was conducive to the escape of bubbles from the molten pool. Compared with the arc–laser hybrid welding (ALHW) mode, the porosity in the joint was reduced. The mechanical properties of the welded joints obtained from LAHW were also better, and the tensile strength could reach 90.3% of that of the base material.

Xu Guoxiang et al. [16,17] of Jiangsu University of Science and Technology established a three-dimensional model that can directly calculate the dynamic behavior of keyholes and the formation of process porosity after considering many factors such as joint geometric characteristics, droplet transition, molten pool flow, and heat source mode, and they studied the formation process of keyhole porosity in laser-MIG hybrid welding of aluminum alloy material. The results showed that the laser beam dissolved the liquid metal in the molten pool on the surface of the test plate to form a keyhole, and the stable keyhole could provide a "channel" for the bubble to escape, which was conducive to the escape of the bubble from the molten pool. The root of the unstable keyhole was stressed unevenly, the wall of the keyhole was easy to collapse, and the liquid metal of the molten pool was backfilled, preventing the escape of bubbles at the bottom of the keyhole. However, the molten pool created during welding flowed clockwise under the action of the MIG arc's force, and the unescaped bubbles at the bottom of the molten pool were carried by the fluid to the rear of the molten pool to form pores.

Simulation technology must also be based on the actual motion of the keyhole in the molten pool to establish a correct three-dimensional model. In addition, due to many factors such as changes in temperature and humidity during welding and the actual power of laser irradiation on the workpiece surface, the results obtained through simulation technology also have certain deviations from the actual results [18,19]. To accurately observe the dynamic behavior of keyholes in the molten pool, Matsunawa and Miyagi et al. [20–22] of Osaka University successfully obtained dynamic images of keyholes in the molten pool during laser welding of an aluminum alloy through X-Ray synchrotron radiation technology. However, the application of X-Ray synchrotron radiation technology is limited by its high cost, complex operation, and unclear image acquisition. Researchers from Hunan University [23–26] designed a "sandwich" welding specimen based on high-speed camera technology to observe the dynamic behavior of keyholes in the molten pool. The specimen is composed of 304 stainless steel and GG17 quartz glass. The laser beam welds the metal along the bonding line, and the dynamic behavior of the keyhole in the welding process is captured by a high-speed camera through the quartz glass.

Therefore, in order to analyze the porosity suppression mechanism of the laser-MIG hybrid welding process for an aluminum alloy, a laser-MIG hybrid welding test of 7075 aluminum alloy was carried out under different defocus quantities, and the porosity of each welded joint was calculated. A high-speed camera test system was designed and built, and the formation position and movement characteristics of keyholes in the molten pool under different defocusing amounts were captured by the "sandwich" method. The dynamic behavior of keyholes in the hybrid welding process was analyzed, and the interrelation between the dynamic behavior of keyholes and stomata in an aluminum alloy during the laser-MIG hybrid welding process was expounded upon.

2. Experimental Procedure

2.1. Hybrid Welding System

The laser-MIG hybrid welding test equipment used in this paper is shown in Figure 1. The fiber laser is a YLS-10000-SS4 laser produced by IPG Company in the United States (New York, NY, USA); it is paired with the corresponding water-cooling device to stably emit a laser with the set power level. The MIG welding machine is the TPS500i-PLUSE welding machine produced by Fronius Company in Wels, Austria, with the corresponding wire-feeding mechanism and a special graphite wire-feeding tube for the aluminum alloy, which can achieve stable MIG welding. An integrated KUKA six-axis robot and three-axis gantry were used to build the laser-MIG hybrid welding test system, which enables precise welding of complex welding paths and allows the integrated equipment to control the laser power, welding current, welding voltage, and welding speed.

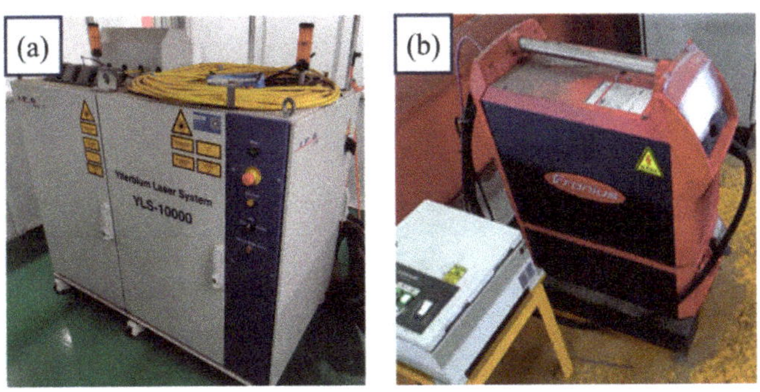

Figure 1. Hybrid welding system: (**a**) fiber laser; (**b**) Fronius welding machine.

The base material for the test was a 7075-T6 aluminum alloy sheet with a thickness of 6 mm, and the welding wire was ER5356. The chemical composition of each material is shown in Table 1. The test employed the butt-welding method, and the test plate specification was $150 \times 100 \times 6$ mm^3.

Table 1. Chemical composition of the 7075 base material and ER5356 welding wire (wt.%).

Element	Cu	Mn	Mg	Zn	Cr	Ti	Si	Fe	Al
7075	1.2~2.0	0.3	2.1~2.9	5.1~6.1	0.18~0.28	0.2	0.4	0.5	Balance
ER5356	0.10	0.4~1.0	4.0~4.9	0.25	≤0.25	0.15	≤0.4	0.40	Balance

Aluminum alloys exposed to the air are prone to an oxidation reaction to produce a dense oxide film; due to the high melting point of the Al_2O_3 film, the weld will develop a large oxide inclusion and non-fusion deficiency. In this experiment, the pretreatment method of mechanical cleaning was adopted. The specific scheme was as follows: First, the edge of the test plate was machined and cleaned with acetone solution to remove the oil on the surface of the test plate. Then, the area to be welded was carefully polished with sandpaper. Finally, the test board was wiped with silk and anhydrous ethanol before welding to remove dust from the surface of the test board. Welding was performed after the anhydrous ethanol was completely volatilized.

A schematic diagram of laser-MIG hybrid welding is shown in Figure 2. In this process, the keyhole shape is greatly affected by the amount of laser defocus. Therefore, the fixed laser power (P) was set to 3.2 kW, the welding speed (v) was 17 mm/s, the welding current (I) was 120 A, and the distance between the laser and the arc (D_{LA}) was 2 mm. The keyhole shape during the welding process was studied by changing the defocusing amount (Δf).

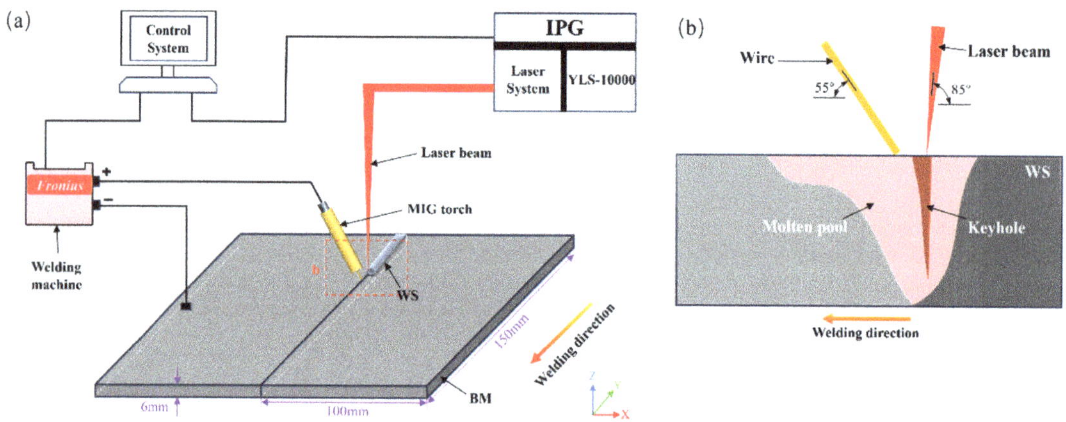

Figure 2. Schematic diagram of laser-MIG hybrid welding. (**a**) Welding system diagram; (**b**) Welding schematic diagram.

2.2. High-Speed Camera System

To observe the change in keyholes' dynamic behavior in the laser-MIG hybrid welding process for an aluminum alloy and study the influence of keyhole shape change on the porosity in the weld, a high-speed camera test system was established as shown in Figure 3. The Acuteye high-speed camera system consists of a camera lens, a CoaXPress high-speed camera, a CoaXPress coaxial cable, PC memory, and a laser-assisted background light source. Because high-speed photography can only capture the opening and closing behavior of the keyhole in the molten pool formed on the workpiece surface in the welding process, the dynamic change in the keyhole's shape inside the weld is unknown, and so this approach provides very limited information on the dynamic behavior of the keyhole and the relationship between the keyhole and the porosity of the weld. Therefore, a new "sandwich" method developed by scholars from Hunan University was selected to observe the dynamic changes in the complete shape of keyholes in the welding process with high-speed camera equipment. The so-called "sandwich" method is to use a piece of quartz glass that has a high light transmission rate and can withstand the action of welding heat instead of the 7075 aluminum alloy base material on one side; the two are closely bonded to form a "sandwich" structure from the aluminum alloy and glass sheet, and the light transmission of quartz glass is used to observe the dynamic behavior of the keyhole inside the weld. The quartz glass selected in this paper is 6 mm thick JGS1, and its physical properties are shown in Table 2. Two-thirds of the laser beam focus was preset on the 7075-aluminum alloy test plate, and the other 1/3 was preset on the JGS1 quartz glass, as shown in Figure 4.

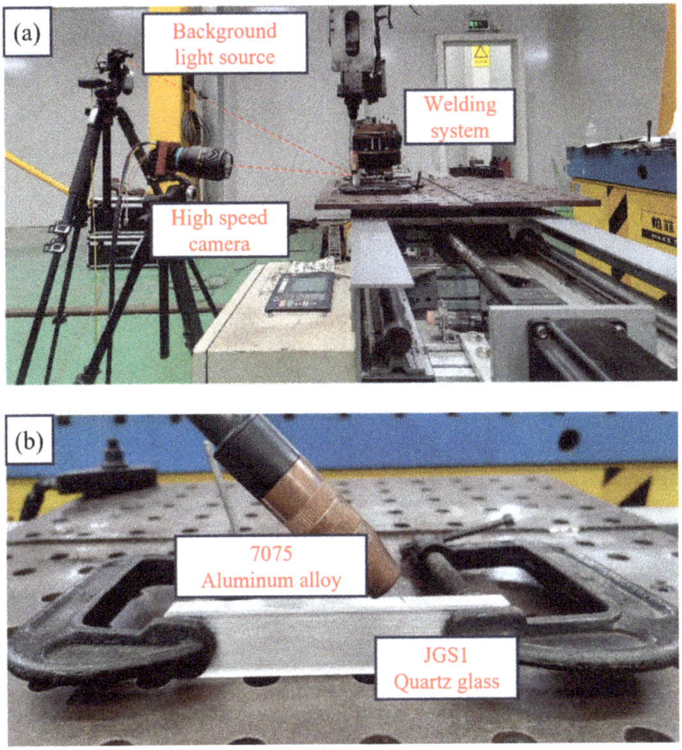

Figure 3. Experimental system: (**a**) system platform; (**b**) local magnification.

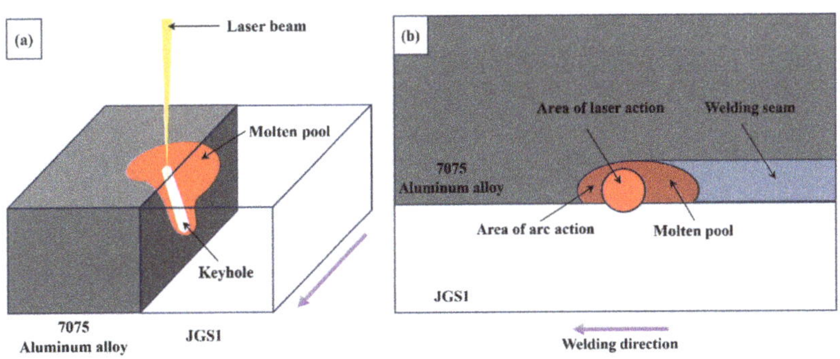

Figure 4. Diagram of the "sandwich" method. (**a**) "Sandwich" method for shooting keyholes; (**b**) preset position of the laser beam focus point.

Table 2. Physical properties of JGS1 quartz glass.

Type No.	Transmittance	Softening Point (°C)	RI	CLTE (K^{-1})	Elastic Modulus (kN/mm^2)
JGS1	92%	1730	1.46	5.5×10^{-7}	48

2.3. Detection and Analysis Equipment

The tensile performance of welded joints was tested using an electronic universal testing machine, as shown in Figure 5a, and the tensile performance test was conducted at room temperature in accordance with GB/T228-2002 "Metallic materials—Tensile testing at ambient temperature" [27]. As shown in Figure 5b, a field emission scanning electron microscope was used to capture the microstructure of various areas in the welded joint, observe the distribution and morphological characteristics of the tensile fracture pores, and determine the types of pores. An XT-3005D X-Ray detector was used for X-Ray detection of weld seams to observe the internal pore distribution of weld seams. The device uses AGFA-C7 film and Lucaid G-39P as the developer.

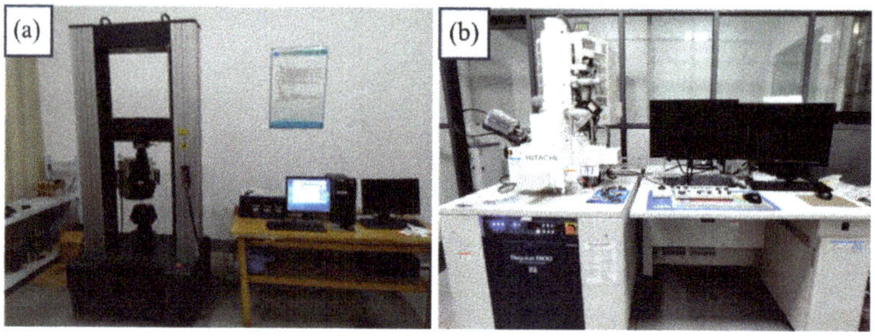

Figure 5. Detection and analysis equipment. (**a**) Electronic universal testing machine; (**b**) field emission scanning electron microscope.

3. Results and Discussion

3.1. Influence of Defocusing Amount on Porosity of Weld

In this paper, a laser power of 3.2 kW, welding speed of 17 mm/s, welding current of 120 A, distance between laser and arc of 2 mm, and defocus of 0 mm were used as the process parameters. X-Ray inspection was performed on the weld made with these parameters, and a large number of pores were found, as shown in Figure 6. To further observe the pore morphology and identify pore types, a tensile sample was taken on the weld, and the position of the tensile fracture was in the center of the weld. There were numerous pores in the fracture, and the porosity was concentrated in the lower part of the weld. The fracture was observed by scanning electron microscopy, and the porosity distribution and morphology of the fracture are shown in Figure 7.

As can be seen from Figure 7, the pore inside the laser-MIG hybrid-welded joint of the aluminum alloy is large in size and irregular in shape; the pores inside are full of folds and not smooth, and there are traces of liquid metal flow. Therefore, it was determined that the porosity is process porosity caused by the instability of keyholes formed by laser beam irradiation in the molten pool. The existence of pore defects will cause the hardness, tensile strength, and corrosion resistance of welded joints to decrease and will seriously affect the service performance of welded joints. The porosity defects formed during laser welding of aluminum alloys have always been an important research topic for domestic and foreign scholars [28,29].

As a high-energy, high-density, and highly penetrating heat source, the laser produces a typical keyhole effect during the welding process. Only a stable keyhole can provide an effective escape channel for bubbles. The pressure from the laser beam irradiation causes the molten pool of metal to arrange itself into a keyhole shape; the keyhole remains stable by balancing the force of gravity on the liquid metal with the force of its surface tension. For the keyhole to maintain a stable state, it must achieve force equilibrium. In laser-MIG hybrid welding, the addition of the MIG arc facilitates the disruption of stress equilibrium in the keyhole, leading to keyhole collapse and the formation of process porosity. Through the literature published by other researchers and our own preliminary tests, we found that the porosity of aluminum alloy laser-MIG composite welding is affected by the influence of welding parameters such as laser power, welding current, and defocusing amount. The main reason for the formation of stomata is the unstable state of keyholes. The change in defocus will affect the shape and stability of the keyhole in the weld pool and thus affect the porosity of the weld. No porosity defect can be found in the weld after welding if the defocusing amount is adjusted appropriately. Therefore, ensuring the stability of keyholes is the key to avoiding the formation of process porosity in aluminum alloy laser-MIG hybrid welding. In this study, the defocusing amount was varied while keeping the laser power, welding speed, welding current, and distance between laser and arc constant. Nine sets of laser-MIG hybrid welding tests were conducted on 7075 aluminum alloy butt plates with different defocusing amounts to investigate the impact of the defocusing amount on welding pores. The test process parameters for the influence of the defocusing amount on the porosity and keyhole morphology of welding seam are shown in Table 3. After welding, X-Ray flaw detection was carried out on the weld corresponding to each defocus quantity, and the distribution of porosity in the weld was observed. After the X-Ray film photography, the local threshold method was adopted to clearly observe the location of porosity distribution in the weld and accurately calculate the proportion of porosity in the weld. After the local threshold method, the film was white as a whole, and the pores were black. For different defocus quantities, the X-Ray films and their processing results are shown in Table 4.

Figure 6. X-Ray film.

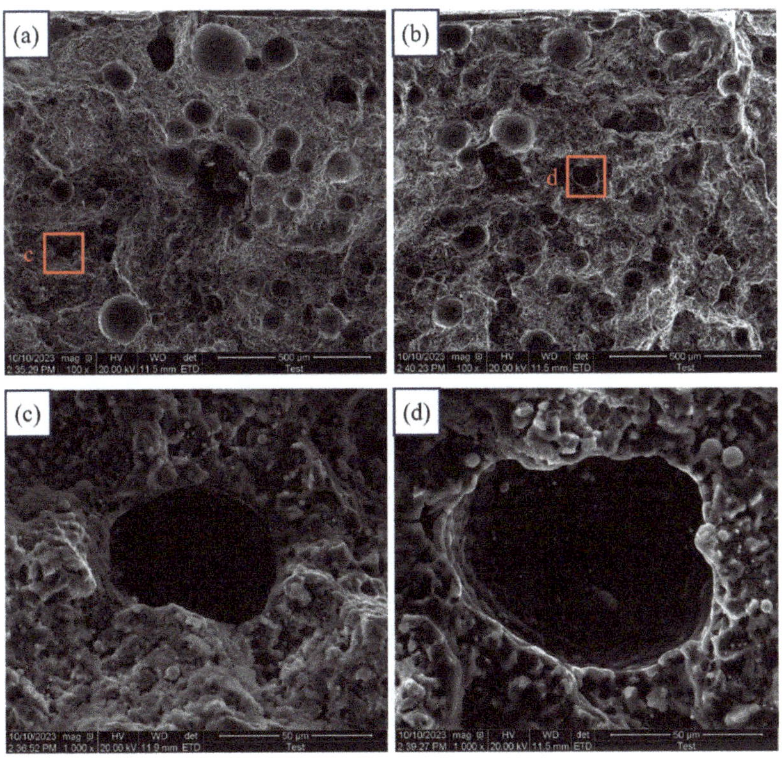

Figure 7. Porosity distribution and morphology on the fracture surface: (**a**,**b**) distribution of pores; (**c**,**d**) morphological characteristics of pores.

Table 3. The defocusing amount affects the tested technological parameters of porosity and keyhole morphology.

No.	Δf (mm)	P (kW)	v (mm/s)	I (A)	D_{LA} (mm)
f-1	+12				
f-2	+9				
f-3	+6				
f-4	+3				
f-5	0	3.2	17	120	2
f-6	−3				
f-7	−6				
f-8	−9				
f-9	−12				

Table 4. Porosity detection results in welds under different defocus conditions.

No.	Δf (mm)	X-Ray Film Detection and Processing Results
f-1	+12	
f-2	+9	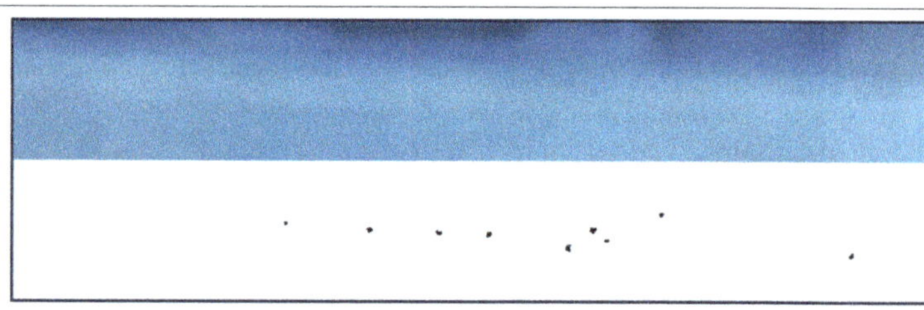
f-3	+6	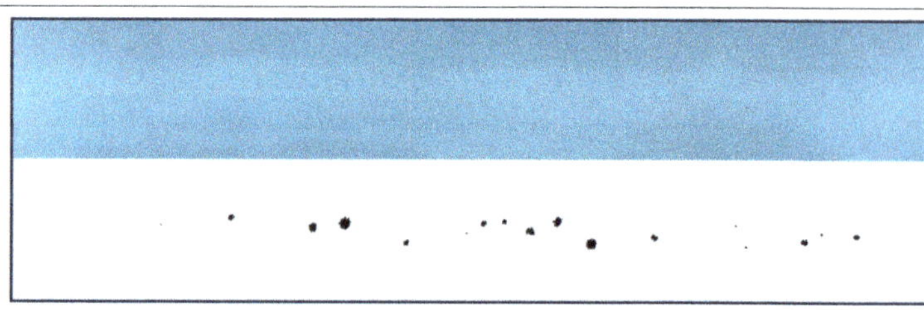
f-4	+3	

Table 4. *Cont.*

No.	Δf (mm)	X-Ray Film Detection and Processing Results
f-5	0	
f-6	−3	
f-7	−6	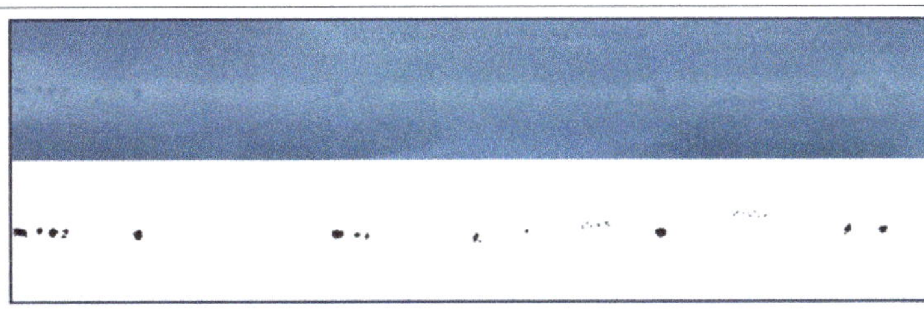
f-8	−9	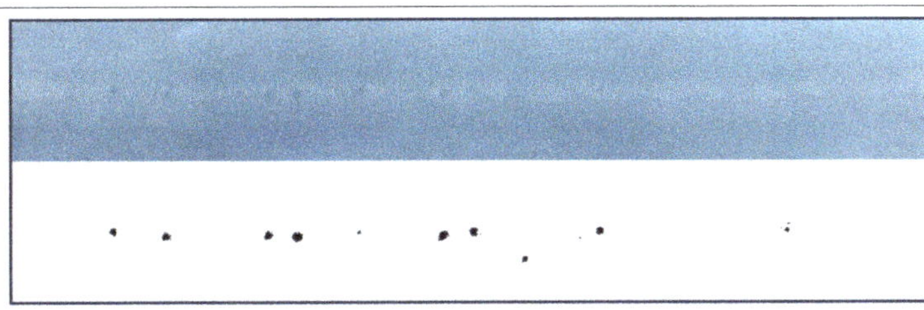
f-9	−12	

Table 4 illustrates that at a defocus of −3 mm, a high density of porosity can be observed in the weld, with both the number and size reaching their maximum. As the defocus shifts from −3 mm towards positive values, there is a noticeable decrease in both the quantity and dimensions of pores within the weld. Notably, at a defocus of +12 mm, no pores are detected in the weld. During testing, when the defocus value exceeds +12 or falls below −12, the quality of weld formation is compromised, resulting in the re-emergence of pores within the welded joint. This occurs because non-optimal defocus—whether too large or too small—leads to a reduction in laser energy density, causing the keyhole to lose its stable state without collapsing and consequently generating a limited number of pores. To accurately quantify porosity within the weld, this study utilizes software to measure pore area on the welded plate and subsequently calculates porosity under varying defocus conditions using Equation (1).

$$n = \frac{\sum S_P}{S_W} \quad (1)$$

In Equation (1), where n represents porosity, S_P denotes the area of a single pore, and S_W stands for weld area. Due to the presence of both the weld and a portion of the base material in the X-Ray test negative image, the value of S_W in this study is defined as half of the area covered by the X-Ray test negative.

The statistical observations on weld porosity under varying defocus conditions are depicted in Figure 8. At a defocus of −3 mm, the weld exhibits maximum porosity at 5.9%. Conversely, at a defocus of +12 mm, no discernible porosity is evident in the weld. As the defocus decreases from −3 mm to −6 mm and then increases to 0 mm, there is a notable reduction in weld porosity to 60% and 45%, respectively; however, it remains relatively high. Furthermore, across nine preset groups of different defocus tests, when the defocus is either −12 mm or +12 mm, resulting in scattered laser beam focus, the porosity in the weld measures less than 1%.

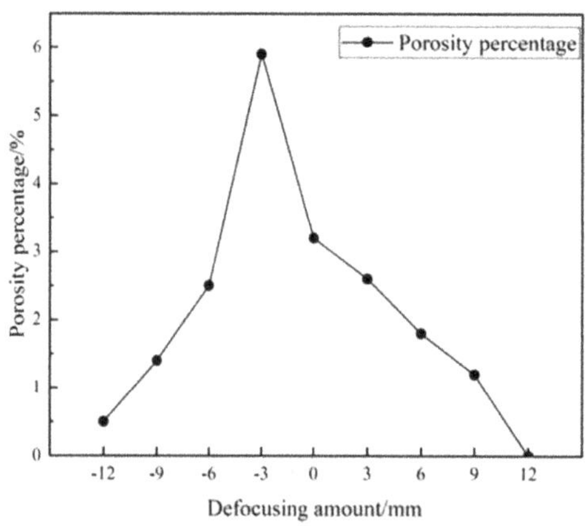

Figure 8. Influence of defocusing amount on porosity.

3.2. *Influence of Defocusing Amount on Keyhole Morphology*

The dynamic changes in keyhole morphology in nine groups of "sandwich"-structured laser-MIG hybrid welds under varying defocusing amounts as listed in Table 3 were captured using high-speed camera equipment. The keyholes' dynamic behavior was categorized into five types: no keyhole formation, collapse of the keyhole root, complete instability of the keyhole, instability of the keyhole root, and stability of the keyhole.

3.2.1. No Keyhole Formed

When the defocus is −3 mm, no discernible keyhole can be observed in the high-speed camera-captured image, as depicted in Figure 9. Among these images, those taken at t + 6 ms, t + 8 ms, and t + 10 ms exhibit slight keyhole formation; however, it is notably unstable. The keyhole collapses at t + 12 ms and does not provide sufficient support for bubble release within its brief duration of existence, leading to the detection of numerous pores in the butt joint.

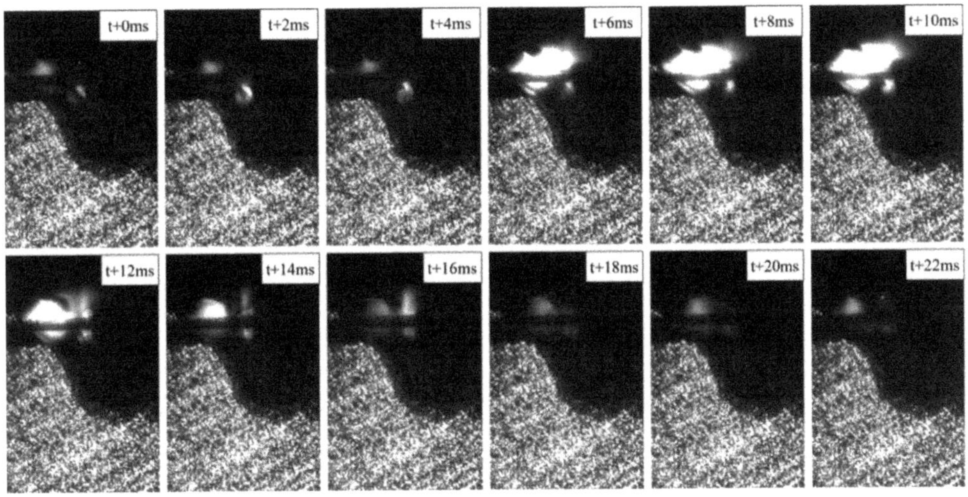

Figure 9. No keyhole is formed when the defocusing amount is −3 mm.

3.2.2. Collapse of Keyhole Root

When the defocus is 0 mm, +3 mm, or −6 mm, the dynamic behavior changes of the keyholes captured by high-speed photography are similar. During the whole welding process, the root keyhole cannot take shape and is always in a collapsing state, while the upper keyhole is always in a stable open state; this situation is called collapse of the keyhole root, as shown in Figure 10. The morphology of the keyhole at a defocus of 0 mm is depicted in Figure 10a. The laser beam is precisely focused on the surface of the test plate, resulting in minimal width and depth of the keyhole. When the defocus is increased to +3 mm, the depth of the keyhole becomes significantly greater, as shown in Figure 10b. When the defocus is reduced to −6 mm, the width of the keyhole becomes significantly greater, as shown in Figure 10c. The collapse of the keyhole at its base can serve as an effective escape pathway for bubbles in the upper region of the molten pool while impeding the release of bubbles in the lower region, thereby accounting for the elevated porosity in welds made under these defocusing conditions.

3.2.3. Complete Instability of the Keyhole

When the defocus increased to +6 mm or decreased to −9 mm, the keyhole shape inside the weld was completely unstable under high-speed photography, as shown in Figure 11. Complete instability of the keyhole means that the keyhole can hardly maintain a stable state during the welding process, and with the movement of the laser beam, the keyhole may completely collapse at any time. When the defocus is +6 mm, as depicted in Figure 11a, the keyhole begins to collapse at t + 4 ms due to sidewall folding, and by t + 10 ms, it has fully collapsed. The condensed metal vapor and protective gas rapidly cool, trapping weld bubbles within the channel and forming a small air mass at the base of the molten pool. The collapsed keyhole begins to be rebuilt until t + 22 ms, during which the small air mass at the bottom of the molten pool is driven by the force of the metal vapor towards the rear of the molten pool to form pores. As shown in Figure 11b, at a defocus of −9 mm, the keyhole collapse process exhibits complete instability; however, the resulting keyhole width and depth are greater, rendering the keyhole relatively more stable. The time of complete keyhole collapse is earlier, ranging from t + 12 ms to t + 16 ms. This observation suggests that the residence time of the keyhole before collapse decreases with an increase in the diameter of the spot illuminated by the laser beam on the test plate surface. Furthermore, it elucidates that the porosity in the weld is lower at a defocus of −9 mm than at −6 mm, as illustrated in Figure 8.

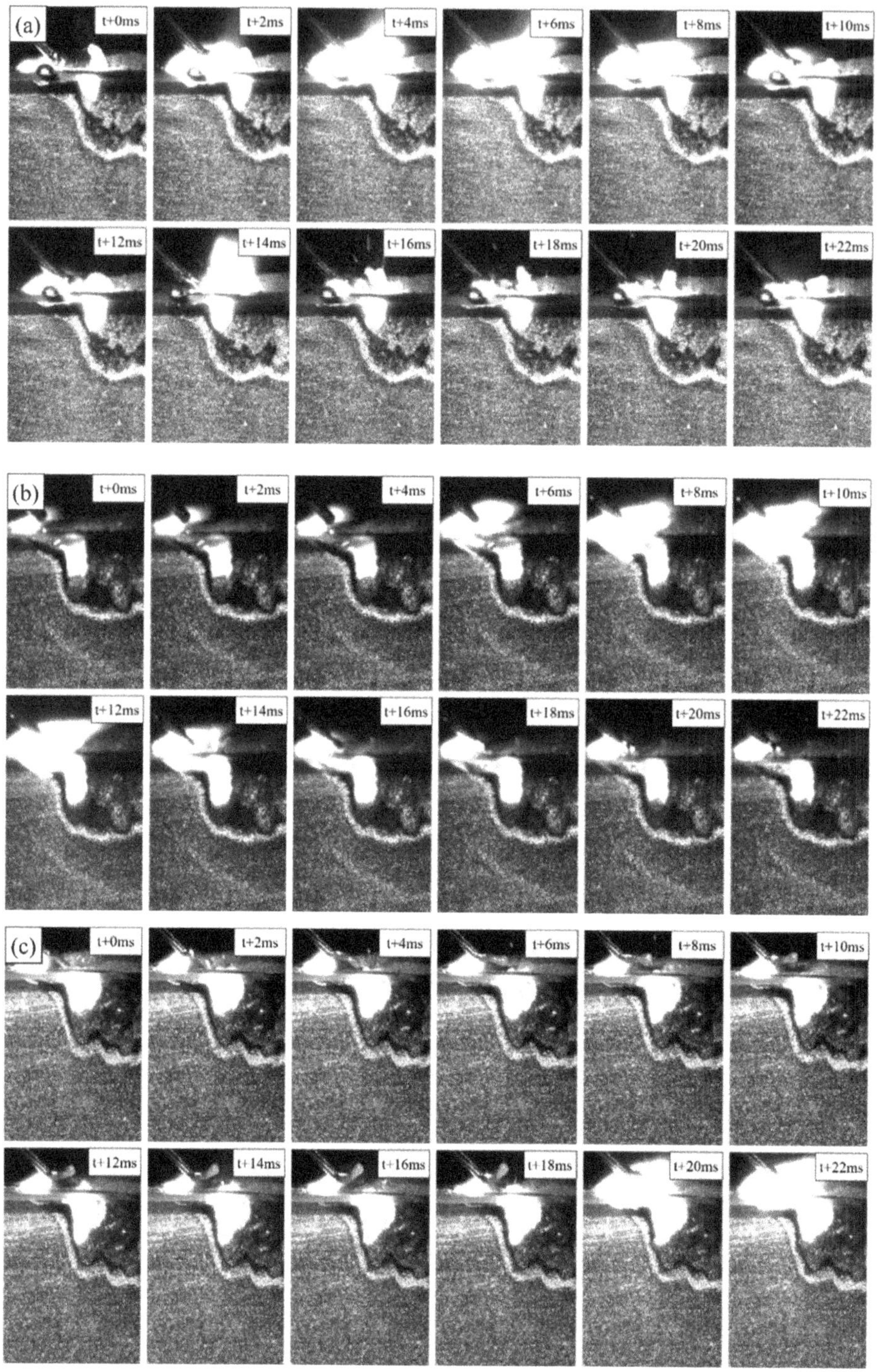

Figure 10. Collapse of the keyhole root. (**a**) The defocus is 0 mm. (**b**) The defocus is +3 mm. (**c**) The defocus is −6 mm.

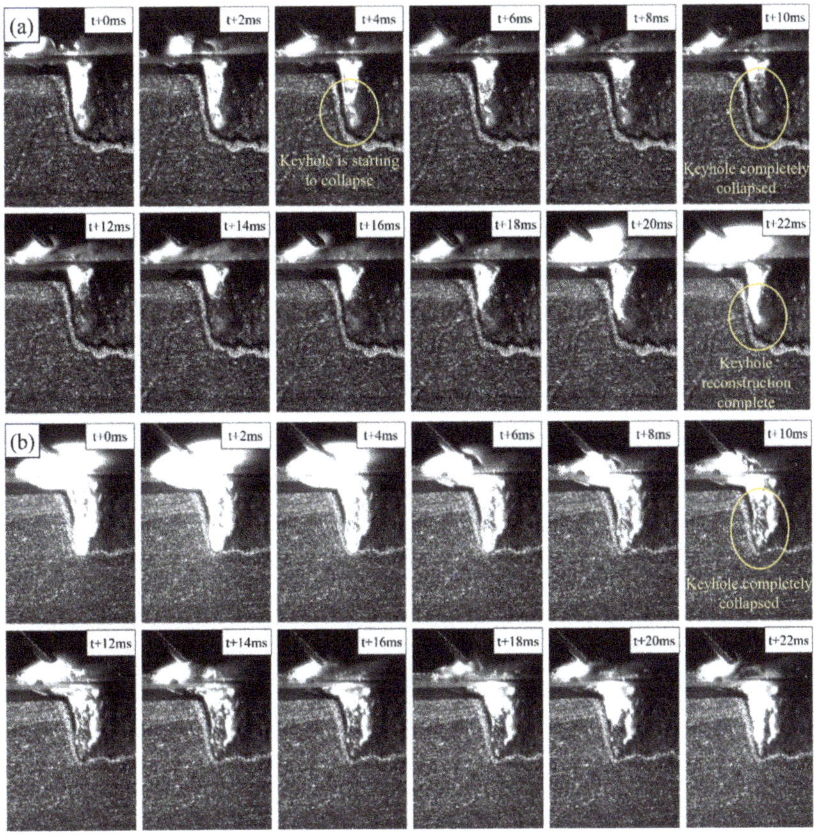

Figure 11. Complete instability of the keyhole. (**a**) The defocus is +6 mm. (**b**) The defocus is −9 mm.

3.2.4. Instability of the Keyhole Root

With a further increase or decrease in defocus, when the defocus is +9 mm or −12 mm, the keyhole morphology is unstable at the root, as shown in Figure 12. In contrast to complete instability of the keyhole, the upper part of the keyhole with the root instability remains in a stable open state; only the lower part of the keyhole collapses with the movement of the laser beam. In contrast to collapse of the keyhole root, the lower part of the unstable keyhole is no longer closed, thus providing a channel for the escape of bubbles at the bottom of the molten pool. Keyholes with root instability neck in the middle, the volumes of trapped metal vapor and protective gas decrease, and the size of the air mass formed at the bottom of the molten pool decreases. With the reconstruction of the keyhole, the number and size of air masses formed by the violent rushing of metal vapor to the rear of the molten pool decreased. The number of pores formed in the weld due to the instability of the keyhole root is small, and the porosity is usually less than 1%.

3.2.5. Stability of the Keyhole

When the defocus is +12 mm, the keyhole is in a stable open state during the entire welding process, as shown in Figure 13. The stable keyhole can provide an escape channel for the bubbles, which is conducive to the escape of the bubbles behind the molten pool. Under this condition, the joint has a good macroscopic shape, and no porosity defects are detected in the weld.

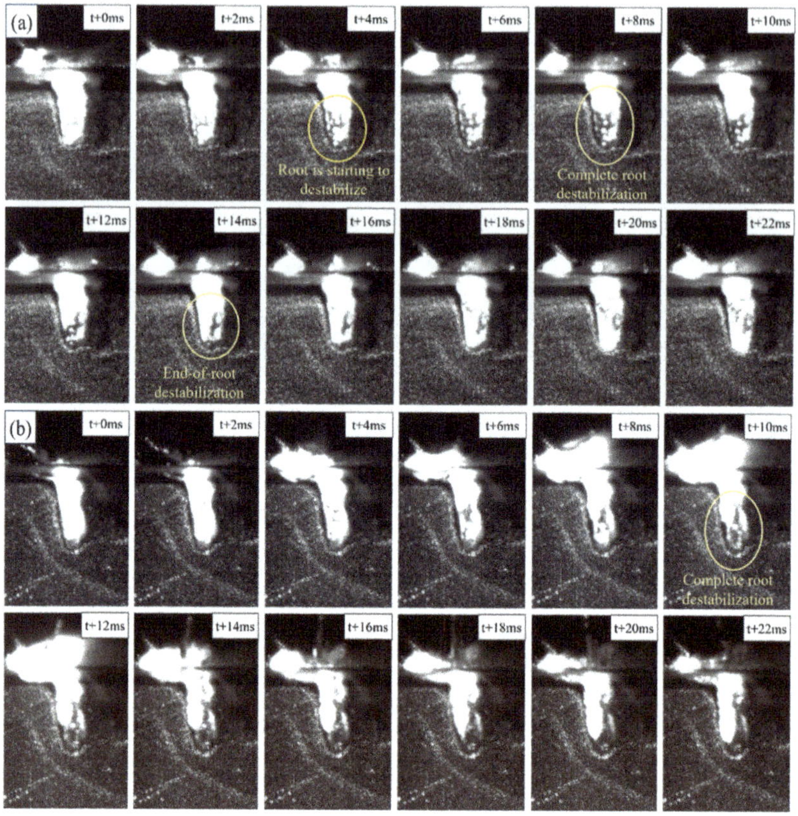

Figure 12. Instability of the keyhole root. (**a**) The defocus is +9 mm. (**b**) The defocus is −12 mm.

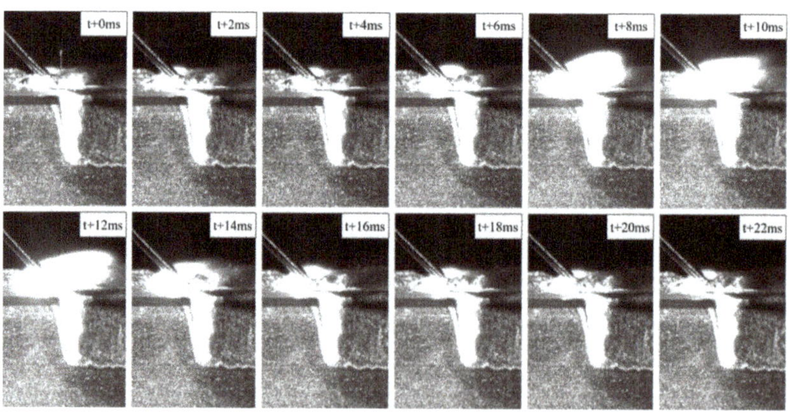

Figure 13. Stability of the keyhole when the defocusing amount is +12 mm.

3.3. Mechanism Whereby the Defocusing Quantity Affects Keyhole Shape

Through the above tests, it is found that the change of the keyhole shape in the weld under different defocusing quantities has a direct effect on the porosity of the weld. During the process of laser-MIG hybrid welding of aluminum alloys, the formation of process porosity is a common occurrence. This can be attributed to the high energy, density, and penetration of the laser as a heat source, which creates stable keyholes in the molten pool. These keyholes play a crucial role in providing effective escape channels for bubbles, as illustrated in Figure 14. From a mechanical point of view, a keyhole comes from the pressure of the laser beam shining on the molten pool to push the molten pool metal to all four sides, but the liquid metal tends to close the keyhole under the action of its own weight, and in order for the keyhole to maintain a stable state, its forces must maintain balance. In laser-MIG hybrid welding, the addition of the MIG arc makes it easier to break the stress balance of the keyhole, and the root of the keyhole or the entire keyhole is unstable, leading to its collapse [30], preventing the escape of bubbles from the molten pool and thus forming pores in the weld.

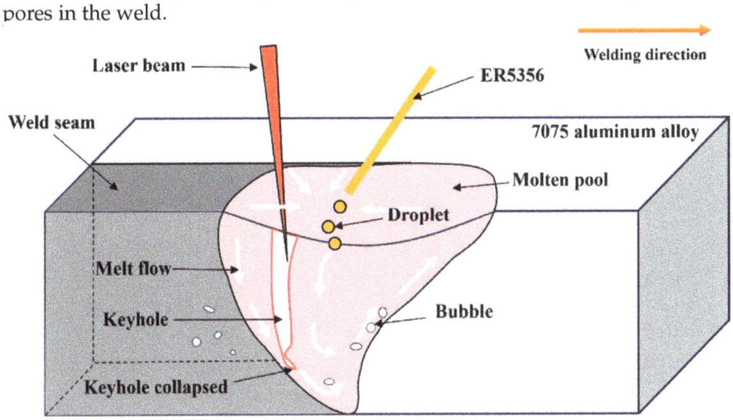

Figure 14. Schematic diagram of the principle of "process porosity" formation.

The main factors affecting the shape of the keyhole are the metal vapor pressure on the inner wall of the keyhole and the pressure of the liquid metal on the outer wall of the keyhole under gravity [31]. To maintain the keyhole in a stable open state and provide an escape channel for bubbles during welding, the metal vapor pressure on the keyhole and the pressure applied by the liquid molten pool under gravity must be balanced [32]. These two forces are only related to the thermal action of the complex heat source, i.e., the laser and the MIG arc, on the molten pool. The thermal action of the keyhole in the molten pool is modeled and analyzed as follows.

The keyhole is placed in a three-dimensional coordinate system represented by (r, φ, z), where r represents the radial coordinate centered on the laser beam, φ represents the azimuth angle, and z represents the axial coordinate. Assuming that there is no heat loss before the laser beam illuminates the molten pool and that all the energy of the laser beam acts on the molten pool formed by the MIG arc, the temperature field distribution in the keyhole of the molten pool formed by laser-MIG hybrid welding can be expressed by Equation (2) [33].

$$T(r,\varphi) = T_L + \frac{P'(r,\varphi)}{2\pi\lambda_{th}} K_0(Pe'r)e^{-Pe'r\cos\varphi} \tag{2}$$

In the equation, T_L represents the metal melting point, λ_{th} denotes the thermal conductivity of the material, P' stands for laser power absorbed per unit depth, K_0 refers to the second class of zero-order Bessel functions, and Pe' signifies the Peclet number.

The Peclet number is a dimensionless parameter associated with welding speed and the thermal physical properties of materials; it can be expressed by Equation (3) [33].

$$Pe' = \frac{v}{2\kappa} \tag{3}$$

In the equation, κ represents the thermal diffusion coefficient of the material.

Because the inner wall of the keyhole is subjected to the heat action of metal vapor and the outer wall is subjected to the heat action of liquid metal, the temperature of the keyhole wall should be between the metal's melting point, T_L, and the metal's boiling, point T_G. Assuming that the temperature of the keyhole wall is raised from the metal melting point to the metal boiling point, $P'^{(r,\varphi)}$ in Equation (2) can be expressed as follows:

$$P'^{(r,\varphi)} = (T_G - T_L)2\pi\lambda_{th}\frac{1}{K_0(Pe\prime r)}e^{Pe' r\cos\varphi} \tag{4}$$

When the laser beam irradiates to the surface of the molten pool, part of the heat is lost due to the reflection of the light, and part of the heat is lost due to the refraction of the light; only the remaining heat will be absorbed by the molten pool. This part of the heat absorbed by the molten pool can be described by the Fresnel absorption power [34,35], and the Fresnel absorption power (Q_F) of the keyhole wall per unit depth can be expressed by Equation (5).

$$Q_F = I2\pi rR(\varphi)\frac{dr}{dz} \tag{5}$$

In Equation (5), I represents the power density of the laser beam, and $R(\varphi)$ denotes the reflectivity of the liquid metal surface in the molten pool to the thermal energy of the laser beam. If the laser power absorbed by the molten pool per unit depth is equivalent to the Fresnel power absorbed by the keyhole wall per unit depth, Equations (4) and (5) can be combined to derive the following:

$$IrR(\varphi)\frac{dr}{dz} = (T_G - T_L)\lambda_{th}\frac{1}{K_0(Pe\prime r)}e^{Pe' r\cos\varphi} \tag{6}$$

In Equation (6), the variable r represents the radius of the keyhole at depth z. When $z = 0$, the value of r denotes the focusing radius of the laser beam. Conversely, when $r = 0$, the variable z signifies the maximum depth of the keyhole. Since the base material used in this test is 7075 aluminum alloy and the welding material is ER5356 aluminum alloy, the parameters representing the thermal physical properties of the materials mentioned above should be constant. In this test, the laser power, welding speed, and laser incidence angle did not change, and the welding process parameters in the above tables should also be constant except for the defocusing amount. By combining all the constants into a single constant C, Equation (6) can be simplified to the following:

$$\frac{dr}{dz} = \frac{Ce^r}{rK_0(r)} \tag{7}$$

According to Equation (7), if the above conditions remain unchanged, the shape of the keyhole is only related to the radius r of the spot where the laser beam reaches the workpiece surface, and the parameter that affects the size of the spot where the laser beam illuminates the workpiece surface is the defocusing amount. The equation explains that the change in the defocusing amount is the fundamental reason for the changes in keyhole shape and porosity in a weld.

4. Conclusions

In this paper, the "sandwich" technique was employed, utilizing quartz glass for lateral observation of the formation position and movement characteristics of the keyhole in the molten pool. The dynamic behavior of the keyhole under nine different defocusing conditions was captured through high-resolution photography, and an analysis was conducted on how varying degrees of defocusing influence keyhole morphology. Furthermore, the porosity of welded joints under varying defocusing levels was quantified and computed to elucidate the intrinsic relationship between keyholes' dynamic behavior and joint porosity. The following conclusions were drawn:

1. The keyhole is formed in the liquid molten pool formed by the laser beam, and the stability of the keyhole directly affects the porosity of the weld. A stable keyhole can provide a "channel" for bubbles to escape, which is conducive to the escape of bubbles from the molten pool. The side wall of an unstable keyhole is stressed unevenly; the root of the keyhole or even the whole keyhole is prone to collapse, and the liquid metal of the molten pool is backfilled, preventing the escape of bubbles from the bottom of the keyhole.

2. With an increase in the positive or negative defocusing amount, the dynamic behavior of the keyhole in the molten pool changes, and the porosity of the weld decreases. When the defocus is −3 mm, no keyhole is formed in the molten pool, and the porosity of the weld is the highest, at 5.9%. When the defocus is 0 mm, +3 mm, or −6 mm, the keyhole root collapses. When the defocus is +6 mm or −9 mm, the keyhole is completely unstable. When the defocus is +9 mm or −12 mm, the keyhole root is unstable. When the defocus is +12 mm, the keyhole is in a stable open state, and the porosity of the weld is 0%.

Author Contributions: Conceptualization, Y.X.; Investigation, Y.X., F.W. and Y.Z; Data curation, Y.X. and Y.Z.; Writing—original draft, Y.X.; Writing—review and editing, F.W., Y.Z., J.F., Z.S. and D.Z.; Supervision, F.W., Y.Z. and J.F.; Funding acquisition, F.W. All authors have read and agreed to the published version of the manuscript.

Funding: This research was funded by the Natural Science Foundation of Jiangsu Province Grant No. BK20241010.

Institutional Review Board Statement: Not applicable.

Informed Consent Statement: Not applicable.

Data Availability Statement: Data are contained within the article.

Acknowledgments: The authors gratefully acknowledge the financial support of the Natural Science Foundation of Jiangsu Province (Grant No. BK20241010).

Conflicts of Interest: The authors declare that they have no known competing financial interests or personal relationships that could have appeared to influence the work reported in this paper.

References

1. Steen, W.M. Arc augmented laser processing of materials. *J. Appl. Phys.* **1980**, *51*, 5636–5641. [CrossRef]
2. Duan, C.F.; Yang, S.L.; Gu, J.X.; Xiong, Q.; Wang, Y. Microstructure and ratcheting behavior of 6061 aluminum alloy laser-MIG hybrid welding joint. *Mater. Res. Express* **2019**, *6*, 086534. [CrossRef]
3. Cai, C.; He, Y.; Xie, J.; Yu, J.; Xu, L.D.; Chen, Z.; Li, Z.X.; Wang, E.H.; Chen, H. Porosity suppression mechanism analysis in narrow-gap oscillating laser-MIG hybrid welding of aluminum alloys based on keyhole stability and molten pool flow behavior. *J. Mater. Res. Technol.* **2024**, *32*, 502–518. [CrossRef]
4. Shibata, K.; Sakamoto, H.; Iwase, T. Laser-MIG hybrid welding of aluminum alloys. *Weld World* **2006**, *50*, 28–34. [CrossRef]
5. Leo, P.; Renna, G.; Casalino, G.; Olabi, A.G. Effect of power distribution on the weld quality during hybrid laser welding of an Al–Mg alloy. *Opt. Laser Technol.* **2015**, *73*, 118–126. [CrossRef]
6. Leo, P.; D'Ostuni, S.; Casalino, G. Hybrid welding of AA5754 annealed alloy: Role of post weld heat treatment on microstructure and mechanical properties. *Mater Design* **2016**, *90*, 777–786. [CrossRef]
7. Sun, Z.; Salminen, A.S.; Moisio, T.J. Quality improvement of laser beam welds by plasma control. *J. Mater. Sci. Lett.* **1993**, *12*, 1131–1133. [CrossRef]
8. Çam, G.; İpekoğlu, G. Recent developments in joining of aluminum alloys. *Int. J. Adv. Manuf. Tech.* **2017**, *91*, 1851–1866. [CrossRef]
9. Jin, X.Z.; Zeng, L.C.; Cheng, Y.Y. Direct observation of keyhole plasma characteristics in deep penetration laser welding of aluminum alloy 6016. *J. Phys. D Appl. Phys.* **2012**, *45*, 245205. [CrossRef]
10. Yan, J.; Gao, M.; Li, G.; Zhang, C.; Zeng, X.Y.; Jiang, M. Microstructure and mechanical properties of laser-MIG hybrid welding of 1420 Al-Li alloy. *Int. J. Adv. Manuf. Tech.* **2013**, *66*, 1467–1473. [CrossRef]
11. Bai, P.C.; Dong, T.S.; Hou, X.H.; Zhao, C.W.; Xing, Y.M. Microstructure and mechanical properties of spray-deposited Mg–12.55 Al–3.33 Zn–0.58 Ca–1Nd alloy. *Mater. Charact.* **2010**, *61*, 756–760.
12. Yan, S.H.; Xing, B.B.; Zhou, H.Y.; Xiao, Y.; Qin, Q.H.; Chen, H. Effect of filling materials on the microstructure and properties of hybrid laser welded Al-Mg-Si alloys joints. *Mater. Charact.* **2018**, *144*, 205–218. [CrossRef]
13. Lin, R.Q.; Wang, H.P.; Lu, F.G.; Solomon, J.; Carlson, B.E. Numerical study of keyhole dynamics and keyhole-induced porosity formation in remote laser welding of Al alloys. *Int. J. Heat Mass. Tran.* **2017**, *108*, 244–256. [CrossRef]
14. Tammas-Williams, S.; Withers, P.J.; Todd, I.; Prangnell, P.B. The influence of porosity on fatigue crack initiation in additively manufactured titanium components. *Sci. Rep.* **2017**, *7*, 7308.

15. Huang, L.J.; Wu, D.S.; Hua, X.M.; Liu, S.C.; Jiang, Z.; Li, F.; Wang, H.; Shi, S.J. Effect of the welding direction on the microstructural characterization in fiber laser-GMAW hybrid welding of 5083 aluminum alloy. *J. Manuf. Process.* **2018**, *31*, 514–522. [CrossRef]
16. Xu, G.X.; Zheng, Z.Q.; Cao, Q.N.; Hu, Q.X.; Li, L.; Guo, Q.H.; Du, B.S. Numerical and experimental investigation on weld formation during laser+ MIG hybrid fillet welding of aluminum alloy in horizontal position. *Int. J. Adv. Manuf. Tech.* **2019**, *102*, 2683–2694. [CrossRef]
17. Xu, G.X.; Li, L.; Wang, H.X.; Li, P.F.; Guo, Q.H.; Hu, Q.X.; Du, B.S. Simulation and experimental studies of keyhole induced porosity in laser-MIG hybrid fillet welding of aluminum alloy in the horizontal position. *Opt. Laser Technol.* **2019**, *119*, 105667. [CrossRef]
18. Li, Y.; Zhao, Y.Q.; Zhou, X.D.; Zhan, X.H. Effect of droplet transition on the dynamic behavior of the keyhole during 6061 aluminum alloy laser-MIG hybrid welding. *Int. J. Adv. Manuf. Tech.* **2022**, *119*, 897–909. [CrossRef]
19. Huang, S.; Xu, L.D.; Lou, M.; Chen, H.; Zhang, K.; Li, Y.B. Keyhole-induced pore formation mechanism in laser-MIG hybrid welding of aluminum alloy based on experiment and multiphase numerical model. *J. Mater. Process. Tech.* **2023**, *314*, 117903. [CrossRef]
20. Matsunawa, A.; Mizutani, M.; Katayama, S. Porosity formation mechanism and its prevention in laser welding. *Weld. Int.* **2003**, *17*, 431–437. [CrossRef]
21. Matsunawa, A.; Kim, J.D.; Seto, N.; Mizutani, M.; Katayama, S. Dynamics of keyhole and molten pool in laser welding. *J. Laser Appl.* **1998**, *10*, 247–254. [CrossRef]
22. Miyagi, M.; Kawahito, Y.; Kawakami, H.; Shoubu, T. Dynamics of solid-liquid interface and porosity formation determined through X-Ray phase-contrast in laser welding of pure Al. *J. Mater. Process. Tech.* **2017**, *250*, 9–15. [CrossRef]
23. Zhang, Y.; Chen, G.Y.; Wei, H.Y.; Zhang, J. A novel "sandwich" method for observation of the keyhole in deep penetration laser welding. *Opt. Lasers Eng.* **2008**, *46*, 133–139. [CrossRef]
24. Zhang, Y.; Li, S.C.; Chen, G.Y.; Zhang, H.R.; Zhang, M.J. Characteristics of zinc behavior during laser welding of zinc "sandwich" sample. *Opt. Laser Technol.* **2012**, *44*, 2340–2346. [CrossRef]
25. Chen, G.Y.; Zhang, M.J.; Zhao, Z.; Zhang, Y.; Li, S.C. Measurements of laser-induced plasma temperature field in deep penetration laser welding. *Opt. Laser Technol.* **2013**, *45*, 551–557. [CrossRef]
26. Jin, X.Z.; Cheng, Y.Y.; Zeng, L.C.; Zou, Y.F.; Zhang, H.G. Multiple Reflections and Fresnel Absorption of Gaussian Laser Beam in an Actual 3D Keyhole during Deep-Penetration Laser Welding. *Int. J. Opt.* **2012**, *1*, 361818. [CrossRef]
27. *GB/T228-2002*; Metallic Materials-Tensile Testing at Ambient temperature. General Administration of Quality Supervision, Inspection and Quarantine of the People's Republic of China: Beijing, China, 2002.
28. Huang, L.J.; Hua, X.M.; Wu, D.S.; Li, F. Numerical study of keyhole instability and porosity formation mechanism in laser welding of aluminum alloy and steel. *J. Mater. Process. Tech.* **2018**, *252*, 421–431. [CrossRef]
29. Zhao, Y.Q.; Zhou, X.D.; Liu, T.; Kang, Y.; Zhan, X.H. Investigate on the porosity morphology and formation mechanism in laser-MIG hybrid welded joint for 5A06 aluminum alloy with Y-shaped groove. *J. Manuf. Process.* **2020**, *57*, 847–856. [CrossRef]
30. Zhu, B.Q.; Zhao, Y.; Chen, F.G.; Fu, J.; Wang, F.Y.; Chen, G.Q.; Qin, Y.H. Mechanism of pore suppression in aluminum alloy laser-MIG hybrid welding based on alternating magnetic field. *J. Laser Appl.* **2024**, *36*, 022016. [CrossRef]
31. Zhang, Z.H.; Dong, S.Y.; Wang, Y.J.; Xu, B.S.; Fang, J.X.; He, P. Study on microstructures and mechanical properties of super narrow gap joints of thick and high strength aluminum alloy plates welded by fiber laser. *Int. J. Adv. Manuf. Tech.* **2016**, *82*, 99–109. [CrossRef]
32. Zhan, X.H.; Zhao, Y.Q.; Liu, Z.M.; Gao, Q.Y.; Bu, H.C. Microstructure and porosity characteristics of 5A06 aluminum alloy joints using laser-MIG hybrid welding. *J. Manuf. Process.* **2018**, *35*, 437–445.
33. Feng, Y.; Gao, X.; Zhang, Y.; Peng, C.; Gui, X.; Sun, Y.; Xiao, X. Simulation and experiment for dynamics of laser welding keyhole and molten pool at different penetration status. *Int. J. Adv. Manuf. Tech.* **2021**, *112*, 2301–2312.
34. Wang, L.; Liu, Y.; Yang, C.G.; Gao, M. Study of porosity suppression in oscillating laser-MIG hybrid welding of AA6082 aluminum alloy. *J. Mater. Process. Tech.* **2021**, *292*, 117053.
35. Salminen, A.; Piili, H.; Purtonen, T. The characteristics of high power fibre laser welding. *Proc. Inst. Mech. Eng. Part C J. Mech.* **2010**, *224*, 1019–1029.

Disclaimer/Publisher's Note: The statements, opinions and data contained in all publications are solely those of the individual author(s) and contributor(s) and not of MDPI and/or the editor(s). MDPI and/or the editor(s) disclaim responsibility for any injury to people or property resulting from any ideas, methods, instructions or products referred to in the content.

coatings

MDPI

Article

Effect of Heat Treatment on Microstructure and Mechanical Properties of Titanium Alloy Fabricated by Laser–Arc Hybrid Additive Manufacturing

Yuhang Chen [1], Juan Fu [1,*], Lilong Zhou [2], Yong Zhao [1], Feiyun Wang [1], Guoqiang Chen [3] and Yonghui Qin [3]

1 Provincial Key Laboratory of Advanced Welding Technology, Jiangsu University of Science and Technology, Zhenjiang 212003, China; nick_cyh@163.com (Y.C.); yongzhao418@just.edu.cn (Y.Z.); wangfeiyunmse@126.com (F.W.)
2 Aerospace Engineering Equipment (Suzhou) Co., Ltd., No. 81 North Guandu Road, Wuzhong District, Suzhou 215104, China; zhoulilongchn@foxmail.com
3 Jiangsu Yangzi-Mitsui Shipbuilding Co., Ltd., Taicang 215400, China; chenguoqiang@yzjship.com (G.C.); qyh@yzjship.com (Y.Q.)
* Correspondence: fujuan@just.edu.cn

Abstract: The tailored thermal heat-treatment process for Ti-6Al-4V alloy manufactured by laser–arc hybrid additive manufacturing can achieve desired microstructures and excellent mechanical properties for components. The effects of different heat treatment regimens on the microstructure and mechanical properties of Ti-6Al-4V alloy manufactured by laser–arc hybrid additive manufacturing are investigated in this study. Utilizing optical microscopy and scanning electron microscopy, we analyze the variations in microstructure with changes in heat-treatment parameters and explore the reasons for the changes in mechanical properties under different solutions' treatment temperatures and cooling rates. The microstructure of Ti-6Al-4V alloy fabricated via laser–arc hybrid additive manufacturing was primarily composed of Widmanstätten α plate structures and a small amount of acicular martensite α' within columnar β grains that grew outward from the substrate along the deposition direction. Following solution treatment and aging heat treatment, the microstructure transitioned to a typical high-performance net basket structure with significantly reduced α plate thickness, leading to noticeable enhancements in sample ductility and toughness. Specifically, when the solution treatment and aging treatment regimen was set at 950 °C for 1 h, followed by air cooling, and then aging at 540 °C for 6 h with subsequent air cooling, the average grain size decreased by a factor of two compared to the as-deposited samples, while the impact toughness increased by 66.7%.

Keywords: laser–arc hybrid additive manufacturing; Ti-6Al-4V alloy; heat treatment; microstructure; mechanical properties

1. Introduction

Ti-6Al-4V alloy, which is known for its excellent performance as a dual-phase titanium alloy, has found widespread applications in a diverse range of fields, including marine vessels, aerospace, and biological research [1]. With the rapid development of manufacturing industries, traditional forming processes struggle to meet the demand for rapid one-time shaping of large structural components [2]. Wire arc additive manufacturing (WAAM) processes provide significant advantages for large component production due to their efficiency, cost-effectiveness, and flexibility [3]. However, WAAM processes suffer from issues such as low forming accuracy and poor material utilization. Laser–arc hybrid additive manufacturing (LAHAM) has garnered considerable attention due to its ability to improve both forming accuracy and production efficiency [4]. Compared to WAAM, LAHAM processes exhibit smaller melt pool sizes and lower temperature gradients at the same scanning speeds, resulting in higher forming accuracy and substantial potential applications [5]

The microstructure of components fabricated using LAHAM is relatively fine, leading to enhanced mechanical properties [6]. In the welding domain, the coupling effect of laser and arc has been shown to stabilize arc behavior while improving the microstructural properties of weld seams. Scholars have begun applying this technique in additive manufacturing. Liu et al. reported that compared to WAAM, LAHAM reduces Zn element vaporization by 5.8%, refines grain size by approximately twofold, and achieves a more uniform distribution of related elements. These improvements resulted in an 11.4% increase in ultimate tensile strength and a 29.9% increase in yield strength [7]. Gao et al. demonstrated that, under the same maximum wall thickness conditions, LAHAM reduces surface roughness by 34.7% compared to WAAM samples, increases effective wall thickness by 20%, and exhibits lower anisotropy [8]. Gong et al. observed that increasing the laser power enhanced the stability of the melt pool during the additive process, but this stability subsequently decreased. Similarly, the surface accuracy of the melt pool initially improved but later deteriorated [9].

It should be noted that the rapid solidification and multiple rapid annealing cycles inherent to the LAHAM process and the microstructure and residual stress distribution of LAHAM Ti-6Al-4V alloy differ significantly from those of forged or cast Ti-6Al-4V alloys [10]. However, there is limited research on the microstructural changes that occur in titanium alloy after additive manufacturing heat treatment [11–15], and our understanding of the microstructural evolution and mechanical properties of LAHAM Ti-6Al-4V alloy, particularly after heat treatment, remains limited, which hinders the development and application of this material. Therefore, this paper primarily investigates the influence of solution treatment parameters such as solution temperature and cooling methods, on the microstructure and mechanical properties of LAHAM Ti-6Al-4V alloy.

2. Materials and Experimental Methods

As illustrated in Figure 1, the LAHAM experimental system was utilized to prepare samples using a LAHAM process. The system comprises an IPG 10kW fiber laser (IPG Photonics Corporation, Newton, Massachusetts, USA), a KUKA six-axis robotic arm, and a Fronius TPS500i single-wire welding machine (Wels, Austria). The arc heat source in this experiment was mainly provided by a TPS500i welding machine manufactured by Fronius. The diameter of the Ti-6Al-4V metal wire (FuShiTe, Baoji, China) was 1.2 mm. During additive manufacturing, the arc current was 178 A, the arc voltage was 21.02 V, the wire feed speed was 6.0 m/min, and the scanning speed was 0.6 m/min. The laser beam was perpendicular to the scanning direction, the included angle between the MIG welding gun and the laser head was 45°, the spot size of the laser at a defocus of 0 was 2 mm, and the distance from the laser to the arc was 2 mm. The experiments were conducted within a controlled atmosphere chamber filled with argon gas. To ensure precision in sample formation due to differences in arc ignition and extinguishing phases, a reciprocating scanning deposition method was employed.

A 12 mm thick Ti-6Al-4V alloy plate was selected as the substrate. Prior to LAHAM processing, the substrate surface was prepared by grinding off the oxide layer using hard alloy files, followed by polishing with sandpaper and cleaning with acetone before LAHAM processing. The LAHAM process parameters were set as follows: laser power, 1500 W, wire feed rate, 6.0 m/min, scanning speed, 0.6 m/min, and argon gas flow rate, 100 L/min. The fabricated LAHAM Ti-6Al-4V alloy samples measured 350 mm × 100 mm × 7 mm.

Microscopic observations of the microstructure were conducted using a ZEISS Axio Observer optical microscope (Oberkochen, Germany), while sample microstructural characterization was performed using a Regulus-8100 scanning electron microscope (ZEISS, Oberkochen, Germany). Tensile tests were conducted at room temperature using a CMT5205 electronic universal tensile (Jinan Precision Testing Equipment Co. Ltd., Jinan, China) testing at a loading rate of 2 mm/min. Two tensile tests were performed under each test condition, and the average value of the test results was taken to ensure the accuracy of the test, while impact tests were carried out using a SANS pendulum impact tester (Jinan,

China). The impact test specimens were prepared in accordance with the standard method of the metal Charpy v-notch impact test. The dimensions of the impact test specimens are shown in the figure. At room temperature, an impact test was performed using a ZBC-2302-D pendulum impact tester (SANS, Jinan, China). The impact energy of the pendulum was 150 J. Three impact tests were performed under each test condition, and the average value of the three tests was taken as the experimental result to ensure the accuracy of the test results.

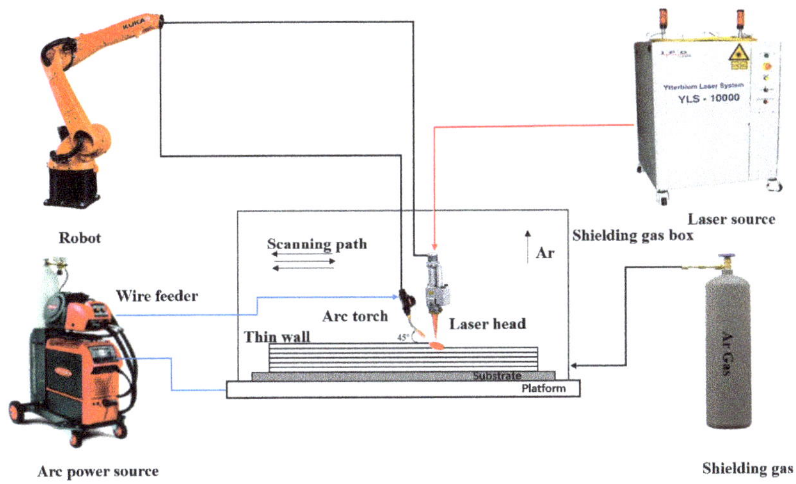

Figure 1. Schematic diagram of LAHAM system.

Additionally, microscopic hardness tests were conducted on thin-walled specimens that had undergone rough grinding, fine grinding, and fine polishing using a German KB fully automatic Vickers hardness tester (KB Prüftechnik Gmb, Lügde, Germany). Microscopic hardness tests were performed at room temperature in the stable region of the additive specimen's cross-section in the vertical direction of the additive. The number of indents for the microscopic hardness test was 50, the spacing between indents was 1 mm, the applied load was 5 N, and the holding time was 10 s. A schematic diagram showing the sampling positions for the tensile specimen, impact specimen, and metallographic specimen are shown in Figure 2. The geometric dimensions of the tensile and impact specimens are shown in Figure 3.

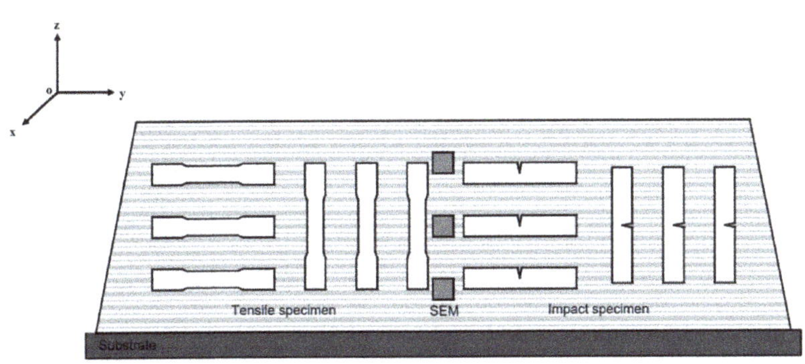

Figure 2. Schematic diagram of the sampling positions for the tensile specimen, impact specimen, and metallographic specimen.

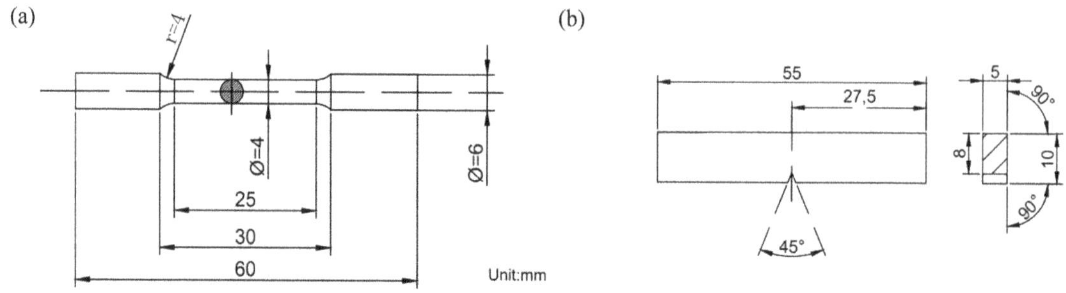

Figure 3. Schematic diagram: (**a**) tensile test specimen; (**b**) impact test specimen dimensions.

The heat treatment regimens and parameters employed for the LAHAM Ti-6Al-4V alloy samples are detailed in Figure 4 and Table 1. During the heat treatment of $\alpha + \beta$ titanium alloy, the size, shape, and distribution of the α phase will affect the β phase. Therefore, the solid solution temperature is generally selected to be within 100 °C below the β phase transition temperature (998 °C ± 100 °C). Based on this, for this experiment, we set the solid solution temperatures to 900 °C, 950 °C, and 1000 °C, and the solid solution time to 1 h. The solid solution process of 950 °C/1 h with relatively good observation and performance was selected to study and discuss the differences in the microstructure and performance of LAHAM Ti-6Al-4V alloy under different cooling methods after "solid solution + aging". Aging treatment can release the lattice strain energy generated after the solid solution, promote the precipitation of the second phase, and decompose the unstable phase. The selection of aging temperature is generally above 500 °C and below the β phase transition point (1000 °C), which can effectively prevent the appearance of the ω phase. As a transition phase between the α phase and β phase, the ω phase has high hardness and brittleness, which will shorten the service life of the workpiece. Based on this, this experiment selected an aging temperature of 540 °C and a holding time of 6h and finally determined the heat treatment process of solid solution + aging at 900 °C, 950 °C, and 1000 °C. To investigate the effects of solution temperatures on microstructure and mechanical properties, the samples were subjected to 1 h solution treatment at 900 °C, 950 °C, and 1000 °C, followed by air cooling. Subsequently, they were maintained at a temperature of 540 °C for 6 h prior to air cooling. To study the influence of the cooling rate on microstructure and mechanical properties, another set of samples underwent 1 h solution treatment at 950 °C followed by furnace cooling, air cooling, and water quenching, These samples were then held at 540 °C for 6 h before air cooling. This experimental design allowed for the elucidation of the interrelationships between the solution temperature, cooling rate, microstructure, and mechanical properties.

Table 1. Heat-treatment parameters for LAHAM of Ti-6Al-4V alloy.

Sample	Heat Treatments *
1	900 °C/1 h, AC + 540 °C/6 h, AC
2	950 °C/1 h, AC + 540 °C/6 h, AC
3	1000 °C/1 h, AC + 540 °C/6 h, AC
4	950 °C/1 h, FC + 540 °C/6 h, AC
5	950 °C/1 h, WQ + 540 °C/6 h, AC

* WQ—Water quenching; FC—furnace cooled; AC—air cooled.

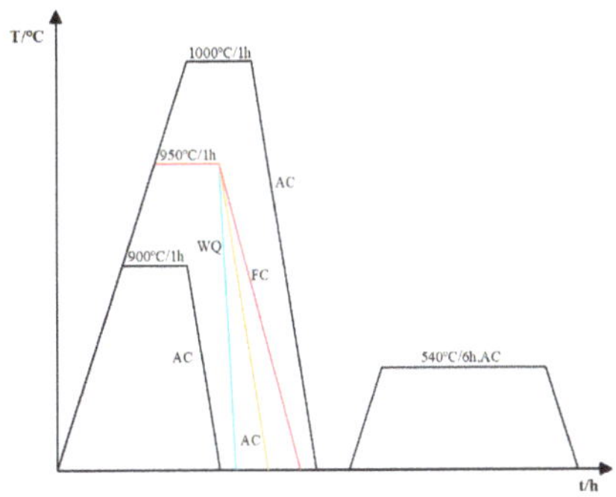

Figure 4. Heat-treatment regimens for solution treatment and aging treatment of LAHAM Ti-6Al-4V alloy samples.

3. Results and Discussion

3.1. As-Deposited Microstructures

Figure 5 shows the surface formation of an as-deposited thin-walled wall. Figure 6a presents the typical macroscopic morphology of LAHAM Ti-6Al-4V alloy, exhibiting continuous extrusion growth of β columnar crystals traversing multiple deposition layers [16]. The average width of the columnar crystals is approximately 350 μm, with α lamellae averaging 2.9 μm in thickness. Due to differing crystal orientations, the macroscopic structure of the deposited walls presents distinct bright and dark banding features. During the LAHAM process, significant temperature gradients develop within the melt pool, perpendicular to the scanning direction. Solidification commences at the bottom of the melt pool, and due to the narrow solidification interval characteristic of Ti-6Al-4V alloy, equiaxed crystal structures struggle to form within the deposition walls. The microstructure within β grains mainly comprises fine Widmanstätten structures, as illustrated in Figure 6b. Bright layering between deposition layers is observed, and this attributed to the reheating effect from subsequent deposition layers, resulting in thermal coarsening of α grains within the heat-affected zone [17].

Figure 5. Surface formation of as-deposited wall.

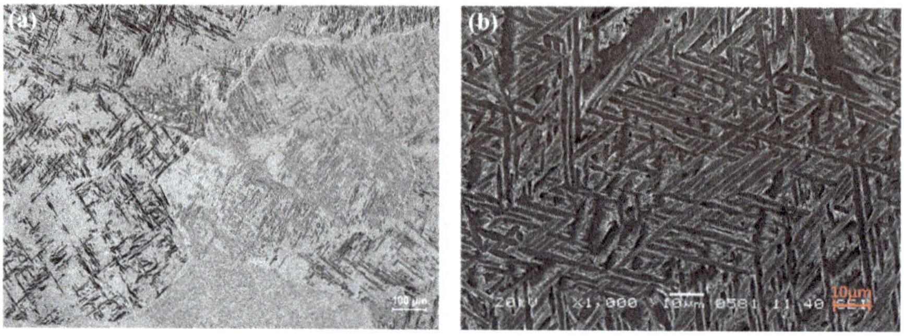

Figure 6. Microscopic structure of as-deposited wall: (**a**) optical microstructure: (**b**) SEM microstructure.

3.2. Role of Heat Treatment in Sample

Solution treatment followed by an aging heat treatment strategy is currently the most commonly used heat treatment strategy for additively manufactured Ti-6Al-4V alloy structural components. Figure 7 illustrates the growth process of the α phase in LAHAM Ti-6Al-4V alloy during the solution treatment process. During this process, the α phase undergoes continuous growth and intersection, causing fragmentation of the primary and secondary α grain and the formation of a denser basket-weave microstructure. The aspect ratio of the α phase significantly influences the mechanical properties of titanium alloys. Complete fragmentation of large primary α grains is crucial for optimizing the mechanical properties of LAHAM Ti-6Al-4V alloy samples. Smaller α grains result in a more uniform microstructure distribution, leading to enhanced overall mechanical properties, particularly in terms of plasticity and toughness.

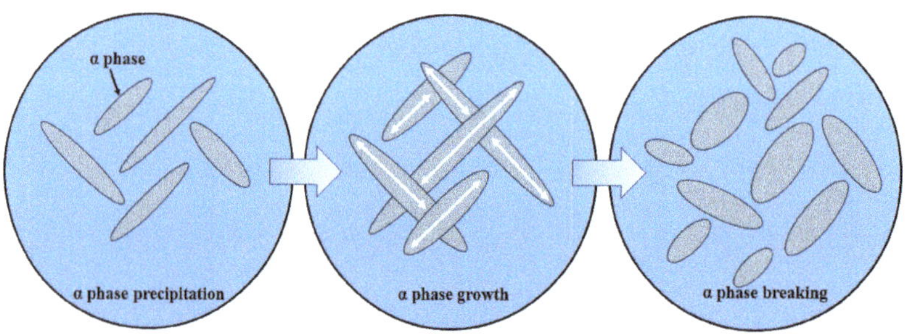

Figure 7. Schematic diagram of the growth process of the α phase in LAHAM Ti-6Al-4V alloy under heat treatment.

Given the similarities in the phases and grain sizes of LAHAM Ti-6Al-4V alloy under different heat-treatment conditions, this study conducted XRD analysis to characterize the samples in the 950 °C/1 h, AC + 540 °C/6 h, AC, and as-deposited states. Figure 8 presents the results of the XRD analysis, revealing significant peak broadening and the presence of the β phase in the LAHAM Ti-6Al-4V alloy samples under the as-deposited state and the 950 °C/1 h, AC + 540 °C/6 h, AC heat-treatment conditions. In combination with the microstructure shown in Figure 6b, it is evident that for the as-deposited components, the quantity and size of the β phase can be considered negligible. After solution treatment at 950 °C, a majority of the α' martensite in the components decomposes, forming an α + β structure. In Figure 8, the β peak (110) of the samples under heat-treatment conditions appears at $2\theta = 38°$, with the β peak under 950 °C/1 h, AC + 540 °C/6 h, AC heat-treatment conditions, the width of β peak under heat-treatment conditions is slightly wider than that of β phase peak under as-deposited conditions. The presence of the β phase in titanium alloys significantly enhances their plasticity, improving their deformability and ductility. Therefore, the plasticity of the LAHAM Ti-6Al-4V alloy components under heat-treatment conditions is notably higher than that of the components in the as-deposited state.

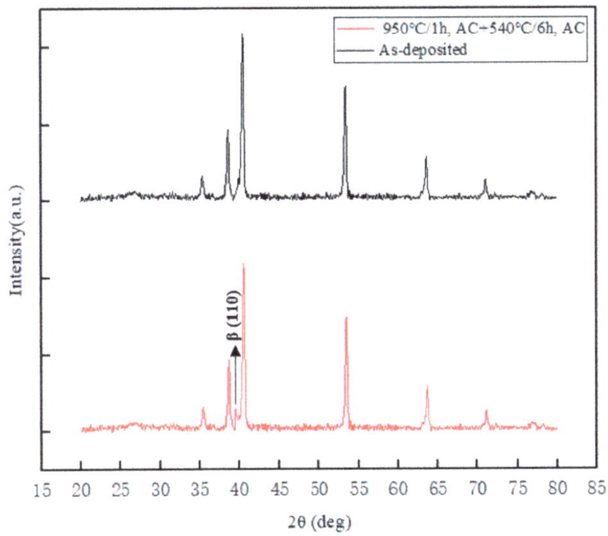

Figure 8. XRD analysis of LAHAM Ti-6Al-4V alloy under heat-treatment conditions.

3.3. *After Heat Treatment Microstructures*

3.3.1. Influence of Solution Temperature

Figure 9a–c present the optical microstructures of LAHAM Ti-6Al-4V alloy after 1 h solution treatment at 900 °C, 950 °C, and 1000 °C, followed by air cooling and 6 h of aging heat treatment at 540 °C. After solution treatment, the microstructure transitions from the typical Widmanstätten structure of the as-deposited state to a basket-weave structure, which consists predominantly of the primary α phase and transformation β phase, wherein the transformation β phase primarily comprises the secondary α phase and β phase [18]. The SEM micrographs of LAHAM Ti-6Al-4V alloy at different solution temperatures are illustrated in Figure 10a–c. From Figure 5, it is observed that as the solution treatment temperature increases from 900 °C, 950 °C to 1000 °C, the thickness of α lamellae increases, while the aspect ratio decreases.

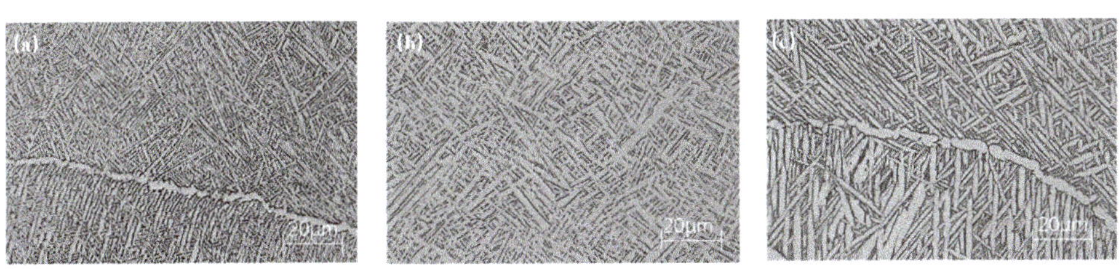

Figure 9. Optical microstructure under different solution treatment temperature conditions: (**a**) 900 °C/1 h, AC + 540 °C/6 h, AC; (**b**) 950 °C/1 h, AC + 540 °C/6 h, AC; (**c**)1000 °C/1 h, AC + 540 °C/6 h, AC.

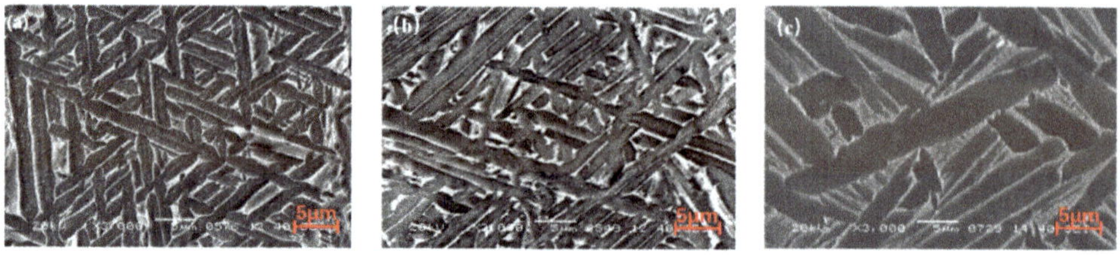

Figure 10. SEM microstructure under different solution treatment temperature conditions: (**a**) 900 °C/1 h, AC + 540 °C/6 h, AC; (**b**) 950 °C/1 h, AC + 540 °C/6 h, AC; (**c**) 1000 °C/1 h, AC + 540 °C/6 h, AC.

At 1000 °C, significant recrystallization occurs within β grains, and due to the precipitation of α phase at β grain boundaries, the previous β grains are no longer continuous. Additionally, Figure 9c presents the β grains; after solution treatment, these transform from columnar to equiaxed form. Figure 11 presents the distribution of α lamellar thicknesses under different heat treatments. The α lath thickness was quantified using Image J software (x64), indicating an increase in α lamellar thickness with the rise in the solution temperature.

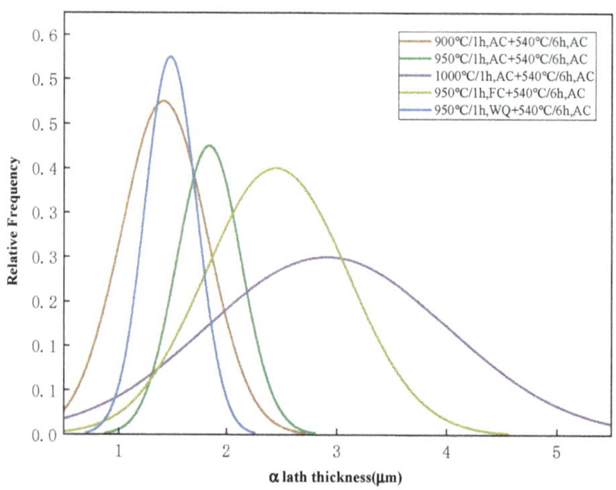

Figure 11. Distribution of alpha lamellar thickness under different heat treatment regimes.

3.3.2. Influence of Cooling Methods

Figure 12a–c display the optical microstructures of LAHAM Ti-6Al-4V alloy after furnace cooling, air cooling, and water quenching to room temperature following 1 h solution treatment at 950 °C, then aging at 540 °C for 6 h before air cooling. SEM microstructures under similar conditions are presented in Figure 13a–c. It is evident from Figures 11 and 12 that as the cooling rates increase, significant changes occur in the microstructure. The thickness of the original α lamellae gradually decreases, while the aspect ratio increases. As depicted in Figure 12a, the slower cooling rate during furnace cooling allows sufficient time for atomic diffusion, providing favorable conditions for the nucleation, growth, and coalescence of the secondary α phase. Consequently, the grain size under furnace-cooling conditions is notably larger than that under air and water-quenching conditions, with partial α lamellae exhibiting spheroidization under slower cooling rates, as shown in Figure 12b. Under air-cooling conditions, a typical basket-weave structure is obtained, wherein the high-temperature β phase transforms into secondary α phase and β phase at an appropriate cooling rate, resulting in reduced α lamellae width compared to furnace-cooling conditions, with adjacent α lamellae oriented differently and interlaced, as illustrated in Figure 13b. Under water-quenching conditions, as observed in Figures 12c and 13c, the α lamellae thickness is significantly smaller than that obtained under the furnace and air-cooling conditions. Moreover, the presence of acicular martensite α' phase with a large aspect ratio in Figure 12c indicates martensitic transformation post-solution treatment, as the rapid cooling rate prevents sufficient atomic diffusion, resulting in the generation of martensite α' phase from high-temperature β phase under faster cooling conditions [19]. The reduction in α lamellae thickness with the increase in the cooling rate observed in the figures will impact the mechanical properties of the samples.

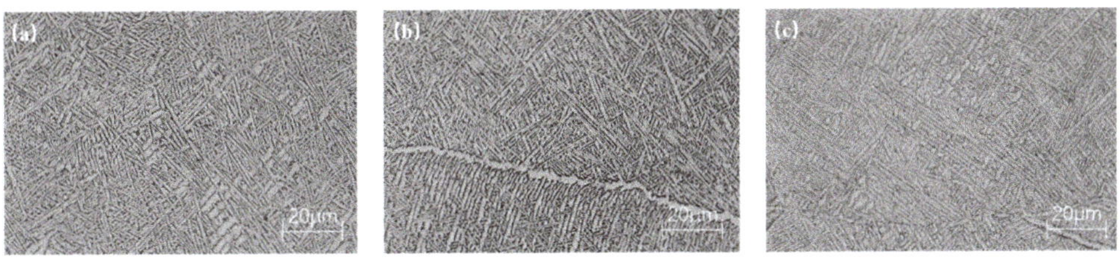

Figure 12. Optical microstructure under different cooling rate conditions: (**a**) FC; (**b**) AC; (**c**) WQ.

Figure 13. SEM microstructure under different cooling rate conditions: (**a**) FC; (**b**) AC; (**c**) WQ.

3.4. Mechanical Properties

3.4.1. As-Deposited Mechanical Properties

The performance of dual-phase titanium alloys is closely related to their chemical composition, phase chemistry, volume fraction, stability, strength, and microstructure [20]. The lower layers of the LAHAM Ti-6Al-4V alloy in the as-deposited state experience multiple thermal cycles, resulting in coarser grains and a quenched microstructure. The average tensile strength is 1100 MPa, the average impact toughness is 15.48 J/cm^2, and the average hardness is 360 ± 25.3 HV.

3.4.2. Different Solution Treatment Temperatures

Figure 14 presents the tensile properties of the LAHAM Ti-6Al-4V alloy after it has been exposed to different solution treatment temperatures. The as-deposited LAHAM Ti-6Al-4V alloy exhibits higher ultimate tensile strength and lower ductility. Upon solution treatment, the specimens exhibit improved ductility, with their strength increasing as the solution treatment temperature rises from 900 °C to 950 °C. However, when the solution treatment temperature increases from 950 °C to 1000 °C, the strength decreases while the ductility continues to decline. Previous studies have shown that the aspect ratio of α laths significantly influences the variations in the tensile strength of Ti-6Al-4V alloys [21]. As the solution treatment temperature rises to below the β transit temperature, the aspect ratio of α laths decreases. Consequently, as more numerous and smaller secondary α phases form, the hindrance to crack propagation increases, leading to improvements in the tensile strength [22]. However, at 1000 °C, the growth of α phases at β grain boundaries disrupts the continuity of the β grain boundaries, reducing their effectiveness in resisting deformation and resulting in decreased tensile strength and elongation. The plasticity of the material is limited by crack nucleation resistance and crack propagation resistance. Larger α laths, which are characteristic of α′ martensite, negatively impact plasticity. Therefore, at a solution treatment temperature of 950 °C, the specimen exhibits a balanced combination of tensile strength and elongation.

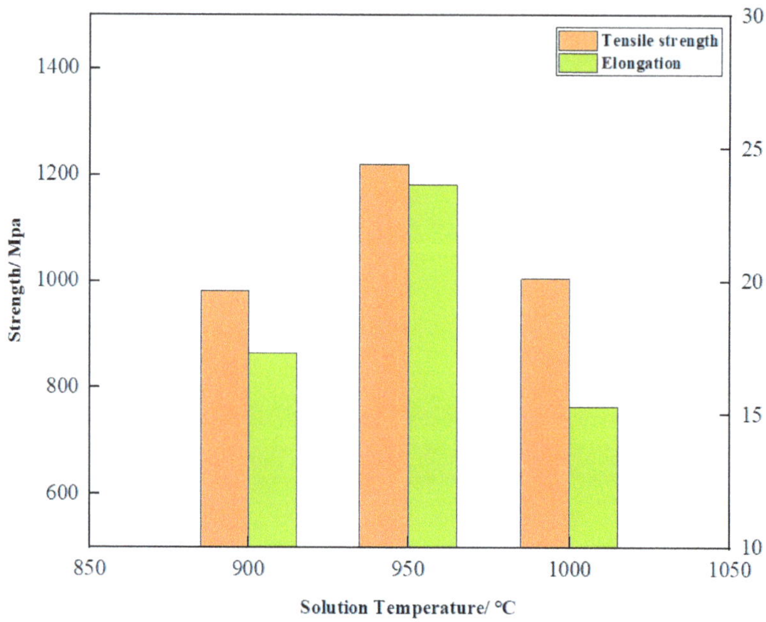

Figure 14. Tensile properties of LAHAM Ti-6Al-4V alloy after solution treatment at 900 °C, 950 °C, and 1000 °C.

In impact experiments, the energy absorbed during sample fracture includes both crack initiation and propagation energies, reflecting the material's ability to resist crack generation and extension [23]. Figure 15 illustrates the impact toughness of LAHAM Ti-6Al-4V alloy at different solution treatment temperatures, and also represents the energy absorbed by the specimen at the fracture points. Corresponding SEM images of impact fracture surfaces are depicted in Figure 16a–c. It is evident that the impact toughness of LAHAM Ti-6Al-4V alloy significantly improves after the solution treatment. This improvement is attributed to the transformation of coarse Widmanstätten structures into interlocking basket-weave structures, where larger orientation differences between adjacent α grains lead to increased resistance to crack propagation [24]. With increasing solution treatment temperature, impact toughness initially rises before declining. At 950 °C, the proliferation of small secondary α phases impedes crack propagation, enhancing impact toughness. However, at 1000 °C, the significant coarsening of the microstructure leads to reduced impact toughness.

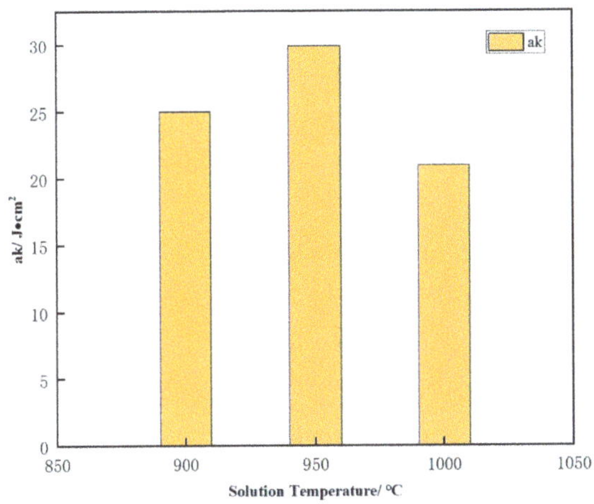

Figure 15. Impact toughness of LAHAM Ti-6Al-4V alloy after solution treatment at 900 °C, 950 °C, and 1000 °C.

Figure 16. SEM morphology of impact fracture surfaces of LAHAM Ti-6Al-4V alloy at different solution treatment temperatures: (a) 900 °C; (b) 950 °C; (c) 1000 °C.

Figure 17 illustrates the hardness of LAHAM Ti-6Al-4V alloy specimens at different solution heat treatment temperatures. As the temperature of the solid solution increases, the hardness of the sample increases and its distribution becomes more uniform. This is attributed to the predominantly interlocked basket-weave structure of the specimens after solution treatment, which is more uniform and less quenched compared to that of the as-deposited Widmanstätten structures. Furthermore, the microhardness of LAHAM Ti-6Al-4V alloy decreases with the increase in the solution heat treatment temperature. At 950 °C, the complete decomposition of α' martensite results in reduced hardness, while at 1000 °C, the coarser α laths and increased grain boundary spacing lead to reduced hardness.

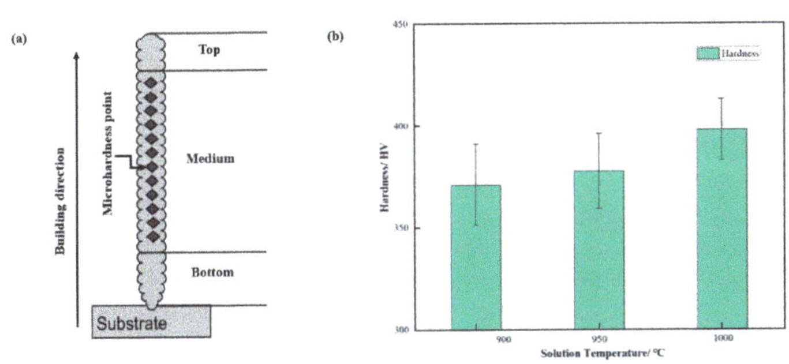

Figure 17. Microhardness measurement: (a) schematic diagram of microhardness position of sample; (b) microhardness of LAHAM Ti-6Al-4V alloy after solution treatment at 900 °C, 950 °C, and 1000 °C.

3.4.3. Different Solution Cooling Methods

Figure 18 presents the room temperature tensile properties of LAHAM Ti-6Al-4V alloy under different cooling rates. It is observed that the ultimate tensile strength of the specimens increases continuously with cooling rates, while the ductility initially increases before decreasing. The specimens subjected to air cooling after solution treatment at 950 °C for 1 h exhibit the best ductility. The grains obtained under furnace-cooling conditions are the coarsest. Larger grain sizes provide more active slip systems, thereby enhancing the ductility [25]. However, excessive grain size reduces the material's ability to deform coherently, leading to reduced ductility [26]. Consequently, the ductility of the specimens under furnace-cooling conditions is superior to that of those under water-cooling conditions but inferior to those under air-cooling conditions. When the grain size is larger, fewer grain boundaries are present to hinder slip. Thus, larger grains have a detrimental effect on strength. The presence of martensite in the water-cooled specimens, with numerous dislocations and twin boundaries, enhances strength but reduces ductility after aging [27]. The α laths in the air-cooled specimens are significantly smaller than those in the furnace-cooled specimens. A smaller grain size enhances the material's ability to resist deformation, resulting in a superior strength–ductility balance in the air-cooled specimens.

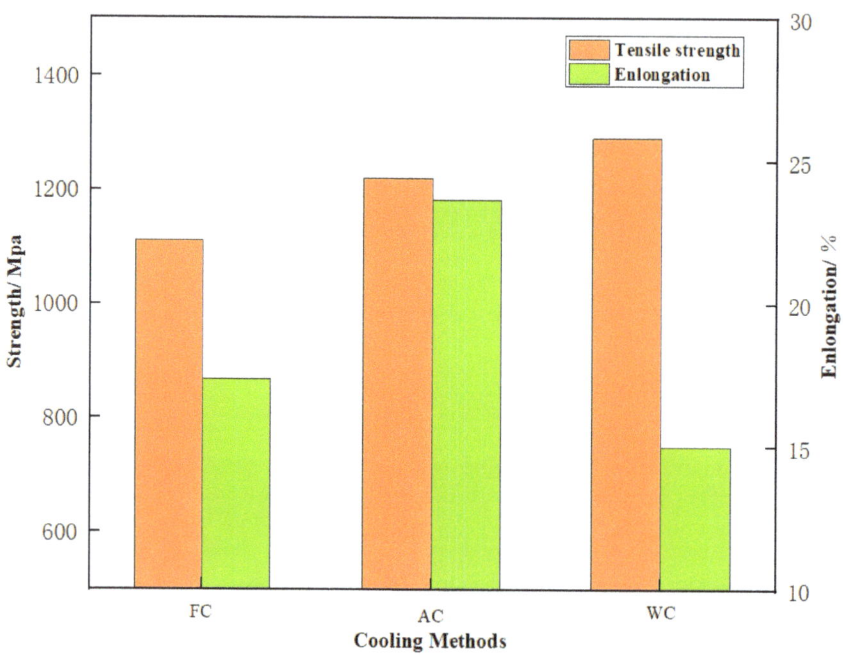

Figure 18. Tensile properties of LAHAM Ti-6Al-4V alloy under FC, AC, and WQ conditions.

Figure 19 presents the impact toughness of LAHAM Ti-6Al-4V alloy subjected to different cooling methods at room temperature. SEM fracture morphology of specimens under different cooling rates is presented in Figure 20a–c. As the cooling rate increases, the impact toughness of the specimens initially increases before decreasing, with the poorest impact toughness observed under water-cooling conditions. Fractures under water-cooling conditions exhibit numerous brittle fracture edges and cleavage steps, indicating poor impact toughness. In contrast, fractures under furnace and air-cooling conditions exhibit numerous irregular dimples, consistent with better toughness of these specimens. From a microstructural perspective, it is reported that during fracture, the α phase serves as the channel for crack generation and propagation in dual-phase titanium alloys, with toughness increasing as the average free path length within the α phase [25,28–31]. The formula for calculating the average free path length within the α phase is provided in Equation (1). Microstructural analysis reveals that the α lath thickness is minimal under water-cooling conditions, with a significant presence of α′ martensite, resulting in the lowest impact toughness. The impact toughness of specimens obtained under furnace-cooling conditions is higher than those obtained under water-cooling conditions but lower than those obtained under air-cooling conditions. This is primarily due to the larger grain size obtained under furnace-cooling conditions, which, according to Equation (1), favors increased impact toughness due to thicker lamellar layers.

$$L_m = \left(\frac{4T_0}{3f}\right)(1-f) \quad (1)$$

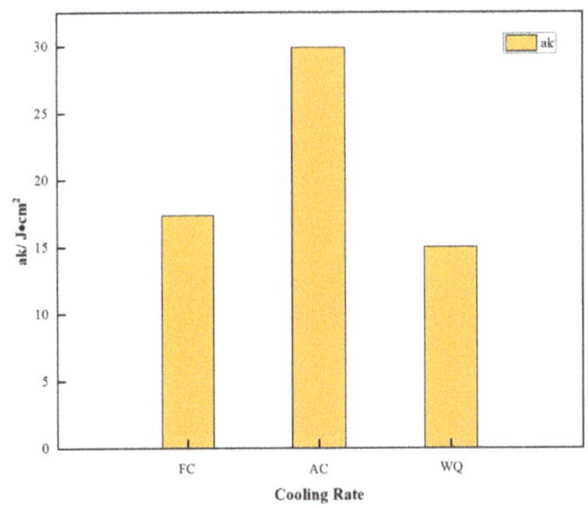

Figure 19. Impact toughness of LAHAM Ti-6Al-4V alloy under FC, AC, and WQ conditions.

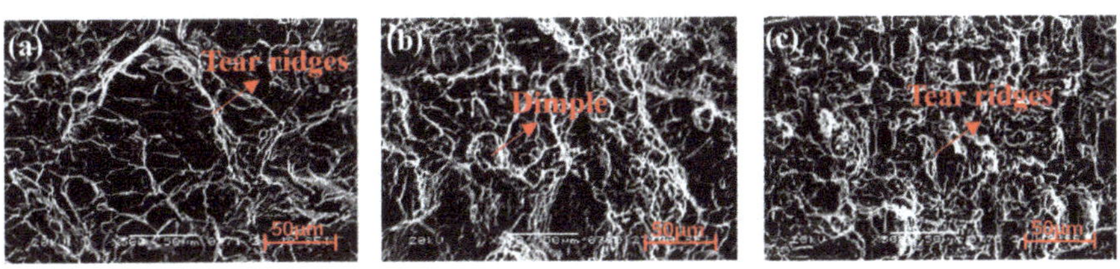

Figure 20. SEM morphology of impact fracture surfaces of LAHAM Ti-6Al-4V alloy under different cooling rates (**a**) FC; (**b**) AC; (**c**) WQ.

However, a larger grain size also reduces the material's ability to deform coherently, resulting in intermediate impact toughness [32]. Under air-cooling conditions, the smaller size of α and β phases results in their interlocking arrangement, with β grain boundaries hindering crack propagation.

Figure 21 depicts microhardness of LAHAM Ti-6Al-4V alloy after different cooling rates. It is observed that the hardness increases with faster cooling rates. The grains obtained under furnace-cooling conditions are coarser, with larger intergranular gaps and greater grain slip freedom, resulting in lower hardness. In contrast, the grains obtained under air-cooling conditions are smaller than those under furnace cooling. Conversely, the microstructure obtained under air-cooling conditions has a smaller grain size compared to the furnace-cooling conditions, with a higher number of grains per unit volume leading to reduced sliding freedom and increased hardness. The highest hardness is achieved under water-quenching conditions due to the martensitic transformation during the solid solution water-quenching process. Martensite contains numerous dislocations that strongly pin the grain boundaries, resulting in the highest hardness being observed in the microstructure under water-quenching conditions.

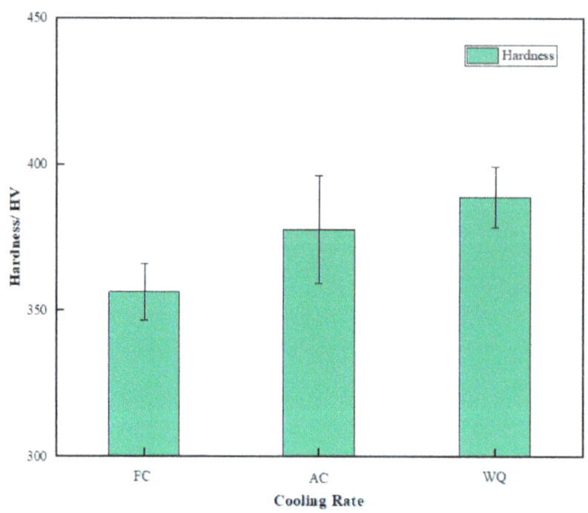

Figure 21. Hardness of LAHAM Ti-6Al-4V alloy under FC, AC, and WQ conditions.

4. Conclusions

This study focused on investigating the microstructural characteristics of LAHAM Ti-6Al-4V alloy after heat treatment. The effects of solution treatment temperature and cooling methods on the microstructure and mechanical properties of the LAHAM Ti-6Al-4V alloy were systematically analyzed.

(1) The microstructural characteristics of LAHAM Ti-6Al-4V alloy consist of columnar β grains with an average grain thickness of 300 μm, which grow epitaxially along the deposition direction. These grains primarily comprise Widmanstätten α laths and a small amount of acicular martensite α'.

(2) As the solution treatment temperature increases, the thickness of primary α laths increases, the aspect ratio of α laths decreases, and the volume fraction of secondary α phases increases. These changes lead to increased strength, decreased ductility, improved impact toughness, and decreased microhardness of the specimens. However, when the solution treatment temperature reaches 1000 °C, severe coarsening of α laths occurs, resulting in simultaneous decreases in strength and ductility. The impact fracture at this temperature exhibits brittle characteristics. The specimens exhibit favorable comprehensive mechanical properties at a solution treatment temperature of 950 °C.

(3) Upon cooling at a sufficiently slow rate, the high-temperature β phase transforms into the low-temperature α phase. At a moderate cooling rate, the high-temperature β phase transforms into plate-like α and β phases. With excessively rapid cooling, the high-temperature β phase undergoes a martensitic transformation, generating martensitic α' phases. As the cooling rate increases, the strength and microhardness of the specimens continuously increase before decreasing. Under water quench conditions, the impact fracture exhibits brittle characteristics. The specimens demonstrate optimal mechanical properties under air-cooling conditions after solution treatment.

(4) During the solution treatment process, α grains grow, intersect with each other, and undergo significant fragmentation, resulting in a noticeable size reduction. Furthermore, the β phase in the LAHAM Ti-6Al-4V alloy increases significantly after solution treatment followed by aging, leading to a notable enhancement in the comprehensive mechanical properties of the samples after heat treatment compared to the components in the as-deposited state. To achieve optimal comprehensive mechanical properties for LAHAM Ti-6Al-4V alloy, a heat-treatment scheme of 950 °C/1 h, AC + 540 °C/6 h, AC is recommended.

Author Contributions: Resources, J.F., L.Z., G.C. and Y.Q.; Writing—original draft, Y.C.; Writing—review & editing, Y.Z. and F.W. All authors have read and agreed to the published version of the manuscript.

Funding: This research received no external funding.

Institutional Review Board Statement: Not applicable.

Informed Consent Statement: Not applicable.

Data Availability Statement: Data are contained within the article.

Conflicts of Interest: Author Lilong Zhou was employed by the company Aerospace Engineering Equipment (Suzhou) Co., Ltd.; Authors Guoqiang Chen and Yonghui Qin were employed by the company Jiangsu Yangzi-Mitsui Shipbuilding Co., Ltd. The remaining authors declare that the research was conducted in the absence of any commercial or financial relationships that could be construed as a potential conflict of interest.

References

1. Wang, W.; Wang, D.; Li, C.; Yang, G.; Ren, Y.; Qin, L. Effect of post heat treatment on microstructure and mechanical properties of Ti-6Al-4V jointing parts proceeded by laser additive manufacturing. *Mater. Sci. Eng. A* **2020**, *788*, 139544. [CrossRef]
2. Fu, J.; Zhao, Y.; Zou, J.; Liu, L.; Pan, Y. Influence of the magnetic field on the melting and solidification behavior of narrow-gap laser welding with filler wire. *Int. J. Adv. Manuf. Technol.* **2022**, *123*, 1123–1131. [CrossRef]
3. Ding, D.; Pan, Z.; Cuiuri, D.; Li, H. Wire-feed additive manufacturing of metal components: Technologies, developments and future interests. *Int. J. Adv. Manuf. Technol.* **2015**, *81*, 465–481. [CrossRef]
4. Näsström, J.; Brückner, F.; Kaplan, A.F.H. Measuring the effects of a laser beam on melt pool fluctuation in arc additive manufacturing. *Rapid Prototy J.* **2018**, *25*, 488–495. [CrossRef]
5. Meng, Y.; Kang, K.; Gao, M.; Zeng, A. Relationship between corrosion resistance and microstructure characteristic of single-pass laser-arc hybrid welded stainless clad steel plate. *Metall. Mater. Trans. A* **2019**, *50*, 2817–2825. [CrossRef]
6. Hann, D.B.; Iammi, J.; Folkes, J. A simple methodology for predicting laser-weld properties from material and laser parameters. *J. Phys. D Appl. Phys.* **2011**, *44*, 445401. [CrossRef]
7. Liu, D.; Wu, D.; Wang, R.; Gan, S.; Niu, F.; Ma, G. Formation mechanism of Al-Zn-Mg-Cu alloy fabricated by laser-arc hybrid additive manufacturing: Microstructure evaluation and mechanical properties. *Addit. Manuf.* **2022**, *50*, 102554.
8. Gao, M.; Zeng, A.; Yan, J.; Hu, Q. Microstructure characteristics of laser–MIG hybrid welded mild steel. *Appl. Surf. Sci.* **2008**, *254*, 5715–5721. [CrossRef]
9. Gong, M.; Meng, Y.; Zhang, S.; Zhang, Y.; Zeng, A.; Gao, M. Laser-arc hybrid additive manufacturing of stainless steel with beam oscillation. *Addit. Manuf.* **2020**, *33*, 101180. [CrossRef]
10. Liu, S.; Shin, Y.C. Additive manufacturing of Ti6Al4V alloy: A review. *Mater. Des.* **2019**, *164*, 107552. [CrossRef]
11. Lee, J.-R.; Lee, M.-S.; Yeon, S.M.; Kang, D.; Jun, T.-S. Influence of heat treatment and loading direction on compressive deformation behaviour of Ti–6Al–4V ELI fabricated by powder bed fusion additive manufacturing. *Mater. Sci. Eng. A* **2022**, *831*, 142258. [CrossRef]
12. Zhang, S.; Lin, X.; Chen, J.; Huang, W. Influence of Heat Treatment on the Microstructure and Properties of Ti-6 Al-4 V Titanium Alloy by Laser Rapid Forming. *Rare Met. Mater. Eng.* **2007**, *36*, 1263–1266.
13. Zhang, G.F.; Leeflang, S.; Böttger, A.J.; Zadpoor, A.A.; Zhou, J. Effect of subtransus heat treatment on the microstructure and mechanical properties of additively manufactured Ti-6Al-4V alloy. *J. Alloys Compd.* **2018**, *735*, 1562–1575. [CrossRef]
14. Vrancken, B.; Thijs, L.; Kruth, J.-P.; Van Humbeeck, J. Heat treatment of Ti6Al4V produced by Selective Laser Melting: Microstructure and mechanical properties. *J. Alloys Compd.* **2012**, *541*, 177–185. [CrossRef]
15. Huang, Q.; Liu, X.; Yang, X.; Zhang, R.; Shen, Z.; Feng, Q. Specific heat treatment of selective laser melted Ti–6Al–4V for biomedical applications. *Front. Mater. Sci.* **2015**, *9*, 373–381. [CrossRef]
16. Carroll, B.E.; Palmer, T.A.; Beese, A.M. Anisotropic tensile behavior of Ti–6Al–4V components fabricated with directed energy deposition additive manufacturing. *Acta Mater.* **2015**, *87*, 309–320. [CrossRef]
17. Thijs, L.; Verhaeghe, F.; Craeghs, T.; Van Humbeeck, J.; Kruth, J.-P. A study of the microstructural evolution during selective laser melting of Ti–6Al–4V. *Acta Mater.* **2010**, *58*, 3303–3312. [CrossRef]
18. Wu, S.Q.; Lu, Y.J.; Gan, Y.L.; Huang, T.T.; Zhao, C.Q.; Lin, J.J.; Guo, S.; Lin, J.X. Microstructural evolution and microhardness of a selective-laser-melted Ti–6Al–4V alloy after post heat treatments. *J. Alloys Compd.* **2016**, *672*, 643–652. [CrossRef]
19. Tao, P.; Zhong, J.; Li, H.; Hu, Q.; Gong, S.; Xu, Q. Microstructure, mechanical properties, and constitutive models for Ti–6Al–4V alloy fabricated by selective laser melting (SLM). *Metals* **2019**, *9*, 447. [CrossRef]
20. Vilaro, T.; Colin, C.; Bartout, J.-D. As-fabricated and heat-treated microstructures of the Ti-6Al-4V alloy processed by selective laser melting. *Metall. Mater. Trans. A* **2011**, *42*, 3190–3199. [CrossRef]
21. Lifshitz, I.M.; Slyozov, V.V. The kinetics of precipitation from supersaturated solid solutions. *J. Phys. Chem. Solids* **1961**, *19*, 35–50. [CrossRef]

22. Bermingham, M.J.; McDonald, S.D.; Dargusch, M.S. Effect of trace lanthanum hexaboride and boron additions on microstructure, tensile properties and anisotropy of Ti-6Al-4V produced by additive manufacturing. *Mater. Sci. Eng. A* **2018**, *719*, 1–11. [CrossRef]
23. Cunningham, R.; Narra, S.P.; Ozturk, T.; Beuth, J.; Rollett, A.D. Evaluating the effect of processing parameters on porosity in electron beam melted Ti-6Al-4V via synchrotron X-ray microtomography. *JOM* **2016**, *68*, 765–771. [CrossRef]
24. Kasperovich, G.; Hausmann, J. Improvement of fatigue resistance and ductility of TiAl6V4 processed by selective laser melting. *J. Mater. Process. Technol.* **2015**, *220*, 202–214. [CrossRef]
25. Ramineni, L.; Almotari, A.; Ali, M.; Algamal, A.; Qattawi, A. Residual Stress Map in Heat-Assisted Additive Manufacturing of IN 718: An X-ray Diffraction Study. *J. Mater. Eng. Perform.* **2024**, *33*, 4124–4135. [CrossRef]
26. Chao, Q.; Hodgson, P.D.; Beladi, H. Microstructure and texture evolution during symmetric and asymmetric rolling of a martensitic Ti-6Al-4V alloy. *Metall. Mater. Trans. A* **2016**, *47*, 531–545. [CrossRef]
27. Brandl, E.; Michailov, V.; Viehweger, B.; Leyens, C. Deposition of Ti–6Al–4V using laser and wire, part I: Microstructural properties of single beads. *Surf. Coat. Technol.* **2011**, *206*, 1120–1129. [CrossRef]
28. Yan, X.; Yin, S.; Chen, C.; Huang, C.; Bolot, R.; Lupoi, R.; Kuang, M.; Ma, W.; Coddet, C.; Liao, H.; et al. Effect of heat treatment on the phase transformation and mechanical properties of Ti6Al4V fabricated by selective laser melting. *J. Alloys Compd.* **2018**, *764*, 1056–1071. [CrossRef]
29. Vasyliv, B.; Kulyk, V.; Duriagina, Z.; Phd, D.M.; Kovbasiuk, T.; Tepla, T. Estimation of the effect of redox treatment on microstructure and tendency to brittle fracture of anode materials of YSZ–NiO (Ni) system. *Восточно-Европейский журнал передовых технологий* **2020**, *6*, 61–71. [CrossRef]
30. Hrabe, N.; Quinn, T. Effects of processing on microstructure and mechanical properties of a titanium alloy (Ti–6Al–4V) fabricated using electron beam melting (EBM), Part 2: Energy input, orientation, and location. *Mater. Sci. Eng. A* **2013**, *573*, 271–277. [CrossRef]
31. Zhao, Z.-Y.; Li, L.; Bai, P.-K.; Wu, L.Y.; Li, J.; Guan, R.-G.; Qu, H.-Q. The heat treatment influence on the microstructure and hardness of TC4 titanium alloy manufactured via selective laser melting. *Materials* **2018**, *11*, 1318. [CrossRef] [PubMed]
32. Baufeld, B.; Brandl, E.; Van der Biest, O. Wire based additive layer manufacturing: Comparison of microstructure and mechanical properties of Ti–6Al–4V components fabricated by laser-beam deposition and shaped metal deposition. *J. Mater. Process. Technol.* **2011**, *211*, 1146–1158. [CrossRef]

Disclaimer/Publisher's Note: The statements, opinions and data contained in all publications are solely those of the individual author(s) and contributor(s) and not of MDPI and/or the editor(s). MDPI and/or the editor(s) disclaim responsibility for any injury to people or property resulting from any ideas, methods, instructions or products referred to in the content.

coatings

MDPI

Article

The Effect of Precipitates on the Stress Rupture Properties of Laser Powder Bed Fusion Inconel 718 Alloy

Jinhong Du [1], Wenhao Cheng [1,2], Yiming Sun [2], Rui Ma [3], Hongbing Liu [1,*], Xiaoguo Song [2,*], Jin Yang [1] and Caiwang Tan [2]

1 School of Materials Engineering, Shanghai University of Engineering Science, Shanghai 210620, China; dujinhong2022@163.com (J.D.)
2 State Key Laboratory of Advanced Welding and Joining, Harbin Institute of Technology, Harbin 150001, China
3 Beijing Power Machinery Institute, Beijing 100074, China
* Correspondence: lhongbing@163.com (H.L.); songxg@hitwh.edu.cn (X.S.)

Abstract: Improving the high-temperature stress rupture properties of Inconel 718 (IN718) alloys is crucial for enhancing aircraft engine performance. By using the laser powder bed fusion (LPBF) technique, IN718 alloys were crafted at varying volumetric energy densities (VED) in this study. The dendrite growth mode, reinforcing phase distribution and high temperature stress rupture properties of various VED samples were investigated. The results showed that the stress rupture life and the uniform elongation of the samples both first increased and then decreased with the increase in VED. When the VED was 60 J/mm^3, the maximum rupture life and elongation of the sample were 43 h and 3.8%, respectively. As the VED increased, the angle of dislocation in the dendrite decreased while the spacing between primary dendrite arms increased, resulting in an increase in the size and volume fraction of the Laves phase. Following a heat treatment, the δ phase would nucleate preferentially around the dissolved Laves phase causing an increase in the volume fraction of the δ phase with the increase in VED. The creep voids readily formed around the δ phase are distributed along the grain boundaries, while the inhomogeneous δ phase and fine grains facilitated crack initiation and propagation. Furthermore, a significant quantity of the δ phase consumed the Nb element, thereby hindering adequate precipitation in the γ'' phase and causing cracks.

Keywords: volume energy density; laser powder bed fusion; Inconel 718 alloy; precipitates; stress rupture properties

1. Introduction

Nickel-based superalloys are gaining popularity owing to their exceptional mechanical properties at high temperatures [1]. Inconel 718 (IN718) is a nickel-based superalloy which finds extensive application in the aviation, aerospace, chemical and energy sectors [2,3]. The alloy exhibits exceptional mechanical properties up to 650 °C, making it well-suited for producing sophisticated components with a high worth and effectiveness, including combustion chamber casings, compressor disks, and turbines [4–6]. Incorporating structural and functional elements into complex components is becoming a crucial means of enhancing performance and innovation for aero-engines and gas turbines [7,8]. However, traditional material processing cannot meet the requirements of integrating lightweight and complex components. Thus, there is a pressing need for a technology that can successfully fulfil the geometric and mechanical requirements of the products.

The integration of structural and functional properties into complex components has gradually become an important approach for innovating and enhancing the performance of aero-engines and gas turbines. However, traditional material processing cannot meet the requirements of integrating lightweight and complex components. Thus, it is imperative to develop a technology that can fulfil the geometric and mechanical specifications of products.

Laser powder bed fusion (LPBF) was developed from the basic principle of prototype manufacturing technology. A three-dimensional digital model of the part was sliced and layered with special software, and the high-energy laser beam was used to selectively melt the metal powder to manufacture 3D solid parts [9–11]. LPBF becomes a key candidate for manufacturing complex IN718 alloy parts due to its cost-effective and time-saving processing features [12].

Many authors have studied the effects of process parameters on the mechanical properties of LPBF IN718 [13–15]. Yang et al. [16] observed that an extension of laser energy resulted in a decrease in tensile power due to the enlargement of the Laves phase. Du et al. [17] studied the effects of build direction on the microstructure and tensile sections of LPBF IN718. It was once found that the fraction of Laves segment and the strength of samples extended with a growing incline angle in the construct direction. The process parameters affected the microstructure and mechanical properties of LPBF IN718 alloy mainly by controlling the Laves phase. However, the Laves phase would be converted to the δ phase after heat treatment, and the δ phase has a greater impact on the high-temperature mechanical properties of the material [18–20].

The presence of the δ phase has been shown, in multiple studies, to alter the microstructure and enhance the mechanical properties of the LPBF IN718 alloy. Zhang et al. [21] studied its high-temperature stress rupture properties and found that samples without an δ phase at grain boundaries had the best stress rupture properties. Shi et al. [22] demonstrated that the heat-treated samples were prone to forming creep voids and microcracks around the δ phase at the grain boundaries during the creep rupture test, thereby accelerating the occurrence of fracture behavior. Gao et al. [20] discovered that an appropriate heat treatment can regulate the morphology and distribution of the δ-phase, leading to an enhancement in the high-temperature mechanical properties of the IN718 alloy. Nonetheless, the latest research has solely examined the impact of different heat treatment methods on the δ phase and mechanical characteristics. The effect of different VEDs on the precipitation behavior in the Laves phase, and the effect of the Laves phase in the as-deposited sample on volume fraction and the distribution of the δ phase after heat treatment deserve further investigation.

The study aims to examine how the Laves phase affects precipitation in the δ phase in samples with varying VED and its impact on stress rupture properties. Additionally, the research aims to establish the relationship between process parameters, precipitates and high-temperature mechanical properties.

2. Materials and Methods

The experiment was conducted using an EOS M290 machine which was fitted with a Yb-fiber laser with a maximum capacity of 400 W. The stainless substrate underwent preheating to reach a temperature of 80 °C. During the printing process, the oxygen content was kept below 100 ppm, and argon was utilized as a protective gas. The chemical composition of the gas atomized IN718 powder is shown in Table 1. The particle size of the IN718 powder ranged from 15 μm to 53 μm, and the scanning electron microscope (SEM) images of the powder are shown in Figure 1. The main parameters are shown in Table 2. The formula for calculating the VED is shown in Equation (1):

$$VED = \frac{P}{v \cdot h \cdot t} \quad (1)$$

where P is the power of the laser, v is the scanning speed, h is the distance between the hatches and t is the thickness of the layer. The scanning strategy was striped, and the layers were rotated by 67°. The as-deposited samples with different VED were named S1, S2 and S3, respectively. The as-deposited samples were heat treated in accordance with the industry standard heat treatment for wrought IN718 alloy [23].

Table 1. The chemical composition of IN718 powder.

Element	Ni	Cr	Nb	Mo	Al	Mn	Si	C	S	Fe
Wt%	52.53	20.19	4.85	3.10	0.5	0.055	0.041	0.025	0.001	Bal.

Figure 1. SEM images characterizing particle size in IN718 powder: (**a**) 1000×, (**b**) 5000×.

Table 2. The parameters of the LPBF process.

Designation	Laser Power (W)	Scanning Velocity (mm/s)	Hatch (mm)	Thickness (mm)	Volume Energy Density (J/mm^3)
S1	210				52.5
S2	240	1000	0.1	0.04	60
S3	300				75

The specific heat treatment was as follows: solution heat treated at 980 °C for 1 h followed by compressed argon cooling to room temperature and 2-stage ageing: 720 °C for 8 h followed by furnace cooling to 620 °C at a cooling rate of 50 °C/h; then the samples were held at 620 °C for 8 h before air cooling to room temperature. After heat treatment, the samples were designated as H-S1, H-S2 and H-S3, respectively.

The metallographic samples prepared by LPBF were initially ground and subsequently polished. When grinding, sandpaper is used in this order (120 mesh, 400 mesh, 800 mesh, 1200 mesh, 2000 mesh, 2500 mesh). After we finished polishing, polishing pastes of W5, W2.5 and W.5 were used for polishing. Afterwards, the polished samples were subjected to etching in a solution containing (5 g $CuCl_2$ + 7 g $FeCl_3$ + 100 mL HCl + 20 mL HNO_3 + 100 of H_2O) for a duration of 20 s. Microstructure investigations were conducted utilizing an optical microscope (OM) and SEM equipment. The Tecnai G2 F30 S-TWIN (FEI Nano Ports Co., Ltd., Waltham, MA, USA) transmission electron microscope (TEM) was employed to identify the distribution of the strengthening phases. TEM samples were prepared using the focused ion beam (FIB) method, and the thickness of the samples was about 100 nm. The Image J software (v1.53) was used to measure strengthening phases size.

Rupture tests were conducted to evaluate the high-temperature mechanical properties of the IN718 alloy. Test samples for the rupture tests were prepared following the ASTM E139 standard [24], as shown in Figure 2. The rupture test was conducted at a temperature of 650 °C and a pressure of 690 MPa on a rupture testing machine RDL100 (Jiangsu Wallong-Hsin Machinery Engineering Corporation Co., Ltd., Wuxi, China). Three samples were tested in each set of experiments.

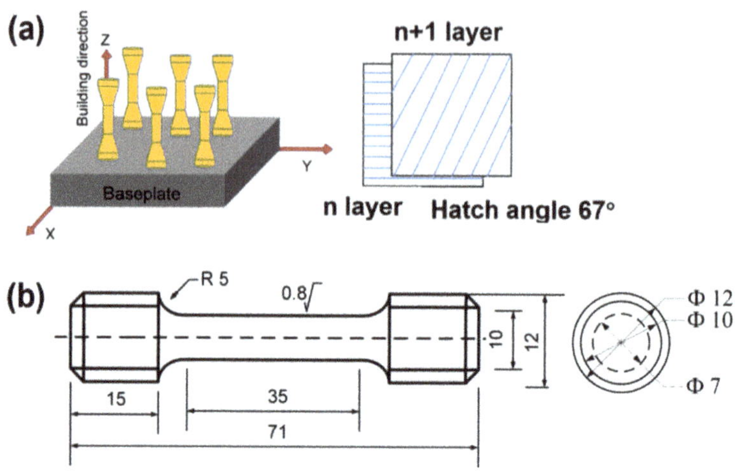

Figure 2. (**a**) Samples of LPBF process, (**b**) Dimensions of stress rupture samples (unit: mm).

3. Results and Discussion

3.1. Microstructure Evolution of LPBF IN718

The microstructure of the as-deposited sample, in the direction of building, is provided in Figure 3. The microstructure of all samples exhibited a characteristic fish-scale appearance along the X-Z direction, which is represented in Figure 3a–c. The columnar grains grew mainly due to the heat flow direction during the solidification of the molten pool being nearly parallel to the deposition direction, and the competitive growth of grains [25]. In as-deposited LPBF IN718 alloy samples, the depth and width of the molten pool increased with greater VED. The length and width of the molten pool were greater for high power laser samples than for low power ones. A higher laser power resulted in grains that were prone to epitaxial growth. Overlapping areas could re-melt and re-solidify during the melting of adjacent tracks or layers due to the high laser power [16,26].

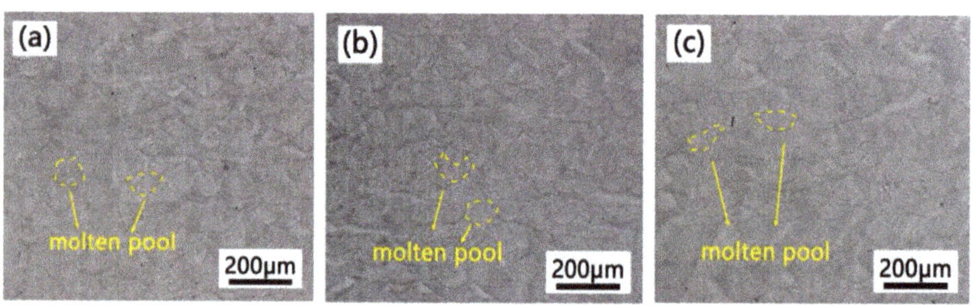

Figure 3. OM images on build direction of the as-deposited IN718 samples: (**a**) S1, (**b**) S2, (**c**) S3.

The segregation of elements (Nb, Mo, Ti) in IN718 alloy leads to the formation of a large number of Laves phases between the molten layers of the alloy [7,27]. The interdendritic region was particularly susceptible to the formation of the Laves phase, as shown in Figure 4a–c. The Laves phase typically measured 1–2 μm in length, and its chemical formula was (Ni, Fe, Cr)2(Nb, Mo, Ti) [28]. The Laves phase would dissolve and transform into the δ phase after heat treatment. The microstructures of the samples post heat treatment are presented in Figure 4d–f. The δ phase was visible in both needle-like and short rod-like shapes. Moreover, the δ phase generally had a size of 0.5–1 μm and a chemical formula of Ni_3Nb in the IN718 alloy [29]. Using Image J software (v1.53), the volume fraction of the Laves and δ phase was determined through statistical analysis (Figure 5). The volume fractions of the Laves phase in different VED samples were calculated to be 6.35%, 7.75% and 9.86%, respectively. After undergoing heat treatment, the respective volume fractions in the δ phase in the three samples were 4.74%, 5.54%, and 7.83%, respectively. The volume fractions of both Laves and δ phases increased with an increase in VED.

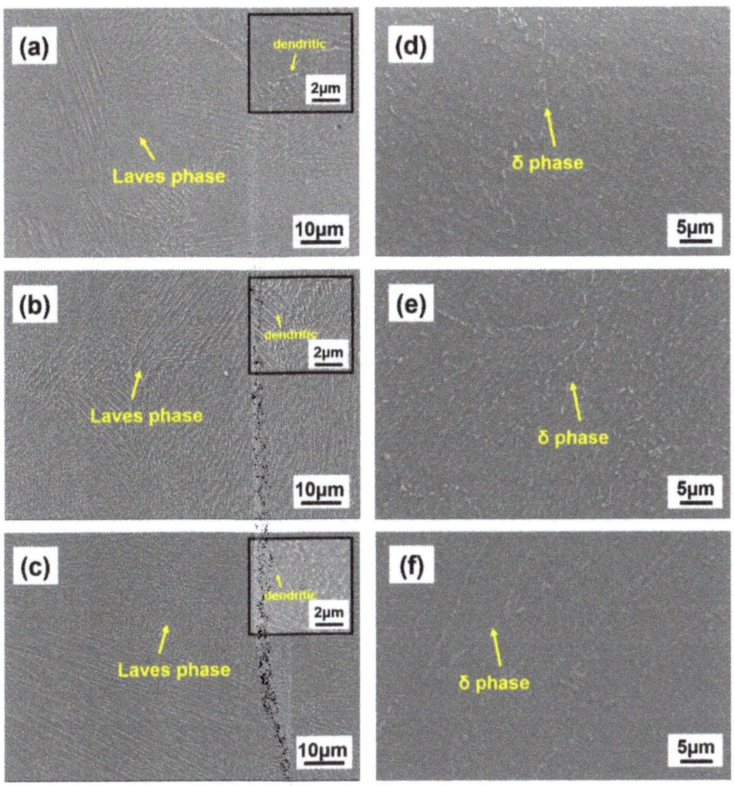

Figure 4. SEM images on build direction of the as-deposited and heat-treated IN718 samples: (**a**) S1, (**b**) S2, (**c**) S3, (**d**) H-S1, (**e**) H-S2, (**f**) H-S3.

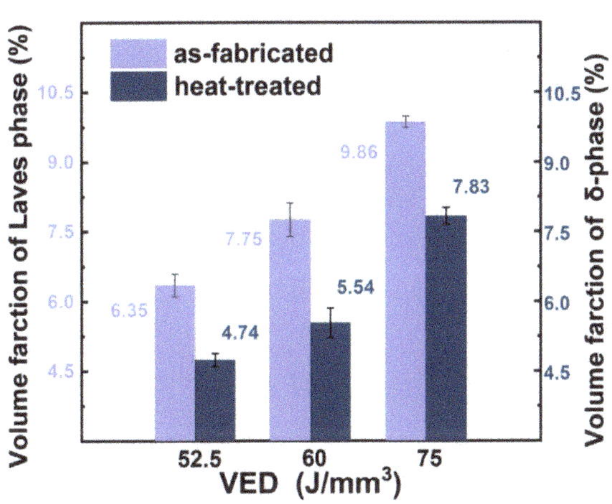

Figure 5. The volume fraction of Laves phase and δ phase in diverse VED samples.

To compare element distribution and precipitate morphologies in the as-deposited and heat-treated samples, we selected S2 and H-S2 for TEM observation. High-angle annular dark field (HAADF) TEM micrographs for the Laves and δ phases appear in Figure 6a,c, respectively. High-density entangled dislocations were found in the intergranular boundaries and matrix regions of S2, in addition to precipitates. The distribution of the main elements can be observed in Figure 6b,d using X-ray energy dispersive spectroscope (XEDS) elemental mapping from the white dotted square. Continual segregation of Nb, Mo and Ti and numerous precipitates along the cellular boundary were observed in the as-deposited sample. However, in H-S2, partial dissolution of the Nb, Mo, and Ti segregation occurred after solution treatment, and the γ'/γ'' phases were precipitated during the aging stage. The entangled dislocations of high density vanished, resulting in a more uniform microstructure.

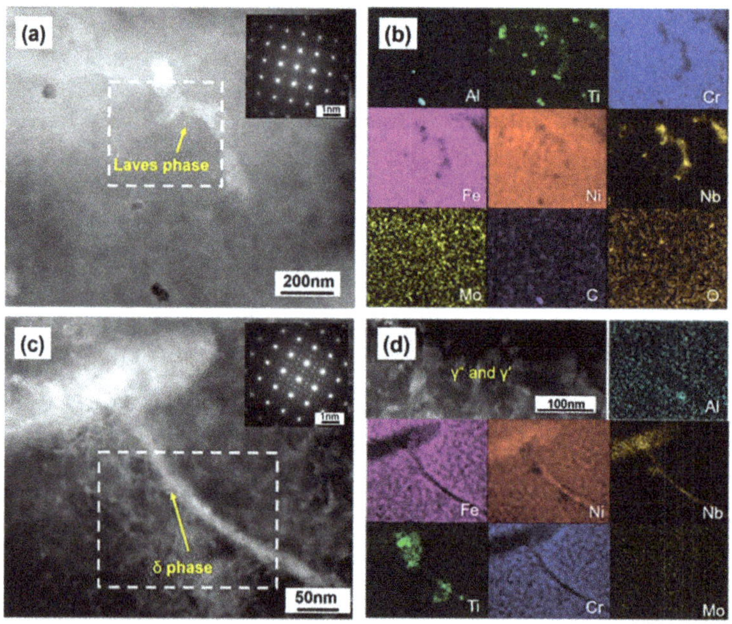

Figure 6. HAADF STEM micrographs and XEDS elemental mapping from the white square: (**a**,**b**) as-deposited sample, (**c**,**d**) heat-treated sample.

3.2. Stress Rupture Properties of LPBF IN718

The stress rupture characteristics of the three sample groups following heat treatment at high temperatures are depicted in Figure 7. The average rupture life of the three samples were 19.5 h, 43.0 h and 32.1 h, respectively. The three samples displayed similar, unsatisfactory elongation rates of only 2.3%, 3.8% and 2.8%, respectively. The average rupture life and elongation of the three samples initially rose before decreasing alongside the increment of VED.

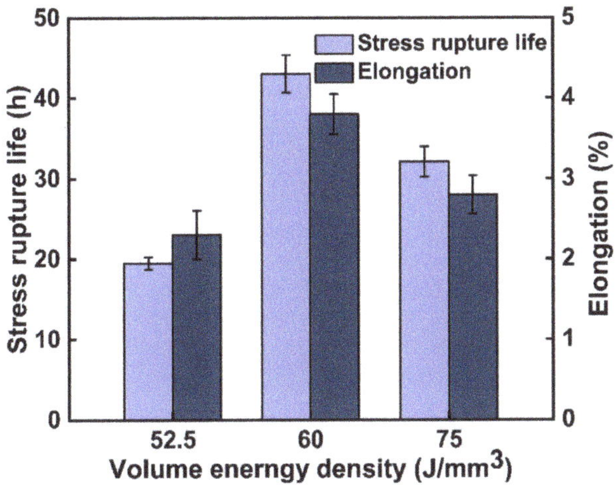

Figure 7. Stress rupture properties of IN 718 at different VEDs.

The fracture surfaces of three samples, as depicted in Figure 8, exhibited intergranular fracture. The mode of fracture was primarily brittle, and the fracture surface was plain, with small and shallow dimples observed in all three samples. As a result, the elongation of the three samples was limited. Micro-voids developed at the interface between the strengthening phase particles and the matrix, facilitating the generation of dimples.

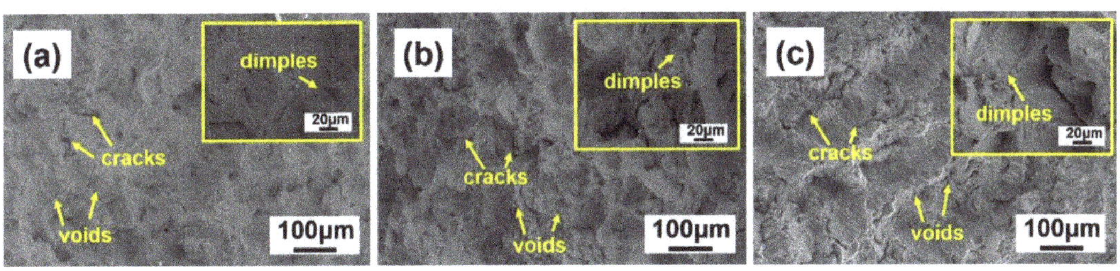

Figure 8. Fractured surface of the samples after creep: (**a**) H-S1, (**b**) H-S2, (**c**) H-S3.

3.3. Analysis of Precipitates in Samples with Different VEDs

To better depict the progression of the Laves phases in the three sample types, the Image J software (v1.53) was employed to tally the lengths, widths and aspect ratios of the Laves phases. Figure 9 demonstrated the statistical findings, whereby the size of the Laves phase grew with increasing VED. The length increased at a slower rate than the average width of the Laves phase, resulting in the aspect ratio of the Laves phase being the smallest in S3 (Figure 9d). The formation of the Laves phase occurred within the interdendritic region during solidification. The morphologies and distributions of the phase were determined primarily by the behavior of dendrite growth [30]. The conventional model for columnar dendrite grain growth suggests that dendrite orientation primarily depends on undercooling differences [31]. For the IN718 alloy, this direction was along its <001> crystallographic orientation, which formed the smallest angle with the local temperature gradient. Each dendrite arm had a misalignment angle θ relative to the temperature gradient ($-45° \leq \theta \leq 45°$ for the 2-D case), as presented in Figure 10a. The V_θ denoted the growth rate of the dendrite arm and was calculated by dividing VL (the velocity of the liquidus isotherm) by the cosine of θ. The cooling rate of molten materials was negatively correlated with the VED, thus a decreasing trend for the misalignment angle θ was shown in S1–S3. In addition, Equation (2) [32] was commonly used to determine the primary dendrite arm spacing (*PDAS*, measured in mm).

$$PDAS = 80\varepsilon^{-0.33} \tag{2}$$

where *PDAS* is the primary dendrite arm spacing of laves phase. ε is cooling rate of molten pool. From the formula, it can be found that *PDAS* was inversely proportional to the cooling rate (ε, K/s). As the VED increased, the samples exhibited a larger PDAS and a smaller misalignment angle θ, as shown in Figure 10b. A small *PDAS* and a large misalignment angle θ easily caused a dense distribution in the Laves phase, as shown in Figure 10c. In contrast, the slower cooling rate at a high VED allowed enough time for Nb, Mo and Ti atoms to diffuse into the interdimer region. This led to the enrichment of abundant atomic groups which served as nucleation sites for the Laves phase. Therefore, more Laves phases were formed at higher VED samples, as shown in Figures 5 and 9.

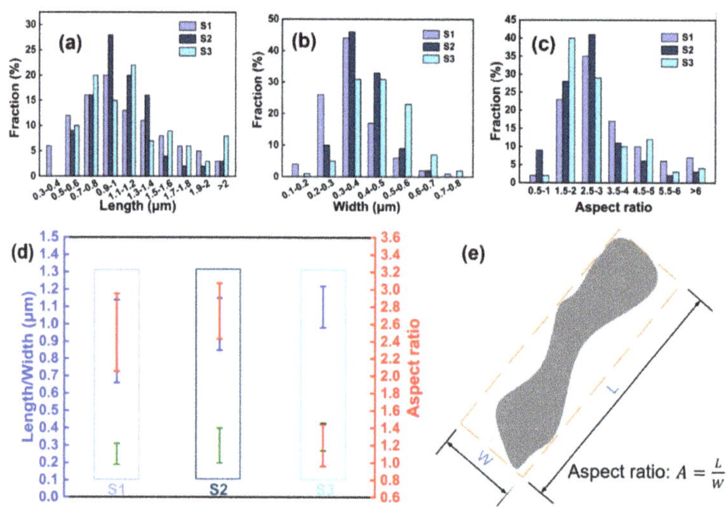

Figure 9. The statistical findings regarding Laves phases in the varied VED samples. (**a**) L (Length), (**b**) W (width), (**c**) aspect ratio, (**d**) the average length, width and aspect ratio of Laves phases, (**e**) the schematic parameter diagram of Laves phase.

The gradual dissolution of the Laves phase consists of three main stages: Laves phase decomposition, solute atom transfer to the γ matrix through boundary and solute atom diffusion [33]. At the beginning of the Laves phase dissolution, three-dimensional diffusion predominantly controlled the process. During the later stage of the dissolution of the Laves phase, the interfacial reaction prevails due to the chemical potential gradient at the interface and the decrease in concentration difference between the Laves phase and the γ matrix, as specified in reference [34]. The δ phase was mainly nucleated at the early stage of heat treatment, so the three-dimensional diffusion mechanism also affected nucleation in the δ phase. The role of Nb in three-dimensional diffusion was significant, and its distribution and diffusion had a considerable impact on the size, morphology and distribution of the δ phase. However, heat treatment of the solution at 980 °C was insufficient to achieve the complete diffusion of Nb. Atoms close to the dissolved Laves phase form diffusion migration first. Therefore, δ phases preferentially produce critical nuclei around the Laves phase [35]. The presence of the Laves phase in the solid solution affected the size and distribution of δ phase in heat treated samples. The contents of Laves phase and δ phase increased with the increase in VED.

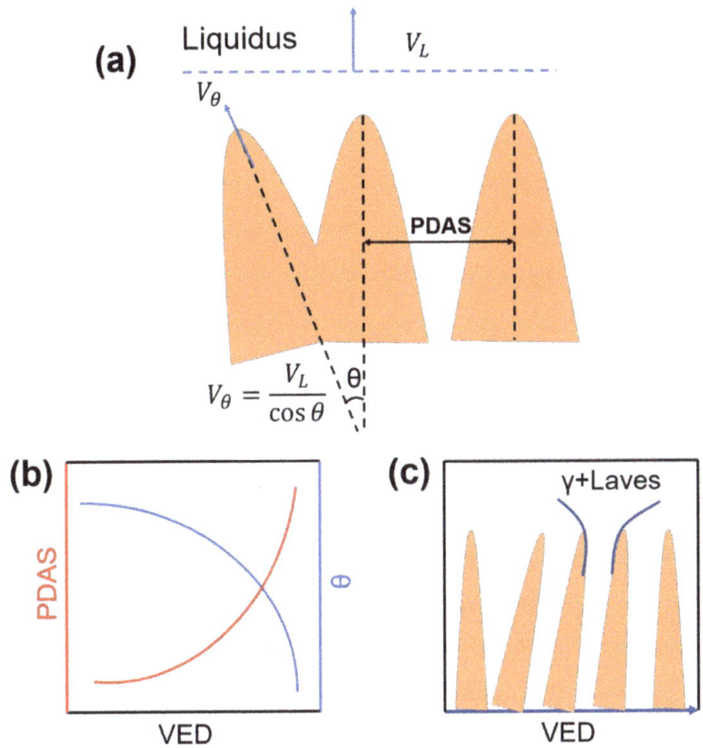

Figure 10. Solidification of columnar dendrites under different VEDs: (**a**) schematic diagram of the competitive grain growth model; (**b**) the relationship between VED and PDAS and misalignment angle θ; (**c**) schematic of solute redistribution under different VED.

3.4. The Influence of Precipitates on Stress Rupture Properties

The longitudinal sections in the rupture fractures were characterized to explain the fracture mechanism, as shown in Figure 11a–c. The plastic deformation ability and stress rupture life of the sample can be indicated by the distribution and length of cracks. The crack length was noticeably shorter in H-S1 than in H-S2 and H-S3. Although the crack distribution in H-S3 was similar to H-S2, the length of cracks was shorter in H-S3 than in H-S2. The cracks primarily occurred at grain boundaries or at the triple point of grain boundaries. The crack number density at approximately 3 mm from the fracture surface of the three sample groups was depicted in Figure 11d. Initially, the samples exhibited a decrease in their crack density as the VED increased. It was found that the crack density is inversely proportional to the durability and elongation of the samples in line with the results of the stress rupture test presented above.

In general, we thought that the long strip of black cracking defects connected together, as shown in Figure 11, and were called micro cracks. Its span size was generally in the range of 1–2 μm. In the creep process, the circular defects that were not connected with the other crack defects were called creep voids. Their sizes were mostly less than 1 μm. Some creep voids would grow up due to tensile stress, as shown in Figure 12b.

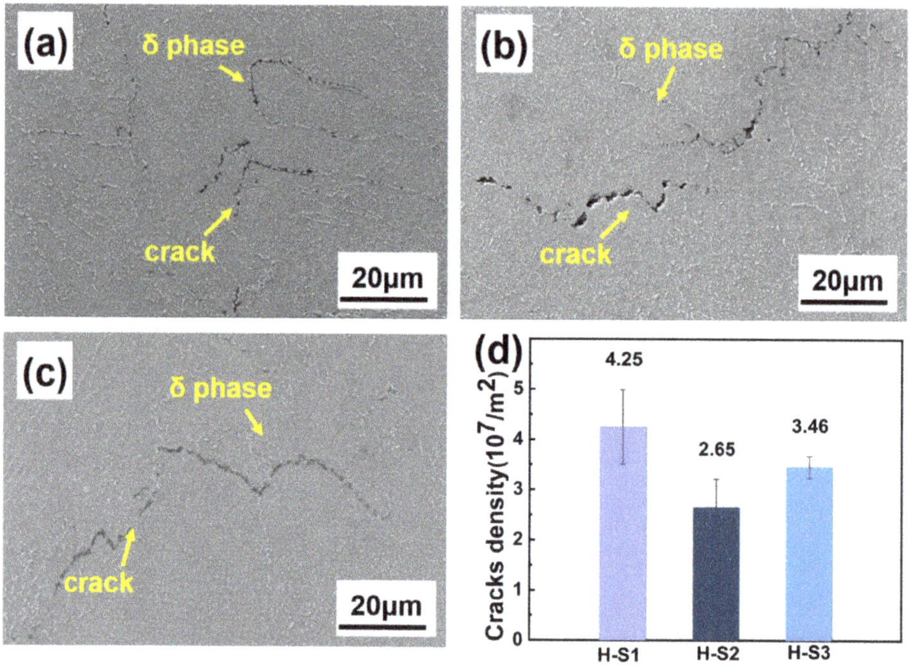

Figure 11. The longitudinal sections of the rupture fractures: (**a**) H-S1, (**b**) H-S2, (**c**) H-S3, (**d**) crack densities of three samples.

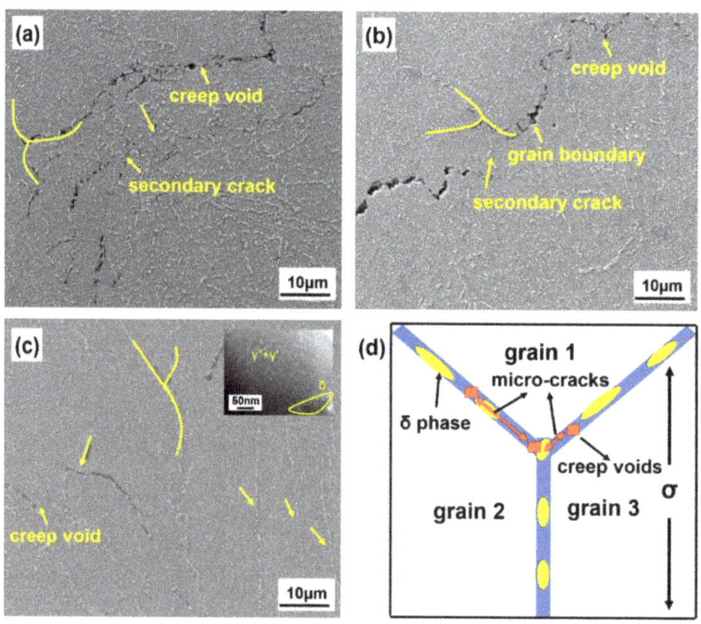

Figure 12. The LPBF IN718 sample after creep: (**a**) H-S1, (**b**) H-S2, and (**c**) H-S3, (**d**) sketch of the creep void and micro-crack formation during creep.

The samples with different VED had an δ phase with different morphologies and sizes after heat treatment. The stress rupture properties were significantly affected by the volume fraction, distribution, and morphologies of the δ phase. As can be seen from Figure 12, the influence of δ-relative durability can be divided into three situations: (1) the δ phase acted as the source of the crack, (2) the crack propagated along the accumulated δ phase, (3) the δ phase affected the precipitation in the γ'' phase, causing cracks. In this investigation, the ratio of the creep temperature (T) to the melting point (T_m) of the IN718 alloy was roughly 0.5. Creep voids had the ability to form and develop at grain boundaries when the T/Tm ratio ranged from 0.3 to 0.6 [19]. Creep voids were found surrounding the δ phase, which was distributed along the grain boundaries, as illustrated in Figure 12a–c. The growth, connection and coalescence of these creep voids resulted in the development of microcracks along the grain boundaries. Furthermore, crack extension could occur at the junction of different grain boundaries due to a disordered atomic arrangement, as shown in Figure 12d.

The S1 had the smallest grain size, and due to the fine-grained material's large grain boundary area per unit volume, there was potential for the nucleation and growth of creep voids at the grain boundaries. The flatter grain boundaries in fine-grained materials made it easier for cracks to propagate. In summary, the H-S1 exhibited the worst stress rupture properties. The uniformly distributed δ phase and suitable grain size in the H-S2 enable it to exhibit optimal stress rupture properties. Groups of δ phases that were aligned parallel to the creep tensile axis were observed in H-S3. These types of δ phases showed minimal impact on the stress rupture properties. However, the δ phase consumed a large amount of the Nb element, so that the γ'' phase could not be sufficiently precipitated in H-S3. Therefore, cracks tended to initiate around a large number of δ phases in the region.

4. Conclusions

In this study, the microstructure evolution and high temperature stress rupture properties of IN718 alloy prepared by LPBF under different VED were analyzed. The following conclusions can be drawn:

1. The existence of the Laves phase in the IN718 alloy was related to the solidification conditions during LPBF process. With an increase in VED, the size and content of Laves phase increased proportionally.
2. After heat treatment, the Laves phase dissolved, resulting in the preferential nucleation of the δ phase in the region where the Laves phase had dissolved in the form of short rods. The volume fraction of δ phase also increased with the increase in VED.
3. Creep voids were easily formed around the δ phase and were distributed along the grain boundary. The inhomogeneous δ phase and fine grains were beneficial to the initiation and propagation of cracks.
4. The δ phase group parallel to the creep tensile axis has little effect on the stress rupture properties. Cracks were more likely to initiate in regions lacking a γ'' phase.

Author Contributions: J.D.: Conceptualization, Investigation, Data curation, Writing—original draft. W.C.: Investigation, Data curation. Y.S.: Conceptualization, Writing—review and editing. R.M.: Investigation, Supervision. H.L.: Funding acquisition, Writing—review and editing, Supervision. X.S.: Resources, Supervision, Project administration. J.Y.: Supervision. C.T.: Writing—review and editing, Supervision, Project administration, Funding acquisition. All authors have read and agreed to the published version of the manuscript.

Funding: The work was supported by the Natural Science Foundation for Excellent Young Scholars of Shangdong Province (ZR2021YQ30) and Zhejiang Key Project of Research and Development Plan (2021C01085) and Key Technology Research Project of Beilun District Ningbo City (2021BLG005).

Data Availability Statement: Data are contained within the article.

Conflicts of Interest: The authors declare no conflict of interest.

References

1. Kim, S.; Choi, H.; Lee, J.; Kim, S. Room and elevated temperature fatigue crack propagation behavior of Inconel 718 alloy fabricated by laser powder bed fusion. *Int. J. Fatigue* **2020**, *140*, 105802. [CrossRef]
2. Pradhan, D.; Mahobia, G.S.; Chattopadhyay, K.; Singh, V. Effect of surface roughness on corrosion behavior of the superalloy IN718 in simulated marine environment. *J. Alloys Compd.* **2018**, *740*, 250–263. [CrossRef]
3. Thakur, D.G.; Ramamoorthy, B.; Vijayaraghavan, L. Study on the machinability characteristics of superalloy Inconel 718 during high speed turning. *Mater. Des.* **2009**, *30*, 1718–1725. [CrossRef]
4. Luo, S.; Huang, W.; Yang, H.; Yang, J.; Wang, Z.; Zeng, X. Microstructural evolution and corrosion behaviors of Inconel 718 alloy produced by selective laser melting following different heat treatments. *Addit. Manuf.* **2019**, *30*, 100875. [CrossRef]
5. Zhao, Z.; Bai, P.; Guan, R.; Murugadoss, V.; Liu, H.; Wang, X.; Guo, Z. Microstructural evolution and mechanical strengthening mechanism of Mg-3Sn-1Mn-1La alloy after heat treatments. *Mater. Sci. Eng. A* **2018**, *734*, 200–209. [CrossRef]
6. Rao, G.A.; Kumar, M.; Srinivas, M.; Sarma, D.S. Effect of standard heat treatment on the microstructure and mechanical properties of hot isostatically pressed superalloy inconel 718. *Mater. Sci. Eng. A* **2003**, *355*, 114–125. [CrossRef]
7. Mantri, S.A.; Dasari, S.; Sharma, A.; Alam, T.; Pantawane, M.V.; Pole, M.; Sharma, S.; Dahotre, N.B.; Banerjee, R.; Banerjee, S. Effect of micro-segregation of alloying elements on the precipitation behaviour in laser surface engineered Alloy 718. *Acta Mater.* **2021**, *210*, 116844. [CrossRef]
8. Zhou, L.; Mehta, A.; McWilliams, B.; Cho, K.; Sohn, Y. Microstructure, precipitates and mechanical properties of powder bed fused inconel 718 before and after heat treatment. *J. Mater. Sci. Technol.* **2019**, *35*, 1153–1164. [CrossRef]
9. Wang, X.; Chou, K. Effects of thermal cycles on the microstructure evolution of Inconel 718 during selective laser melting process. *Addit. Manuf.* **2017**, *18*, 1–14. [CrossRef]
10. Harrison, N.J.; Todd, I.; Mumtaz, K. Reduction of micro-cracking in nickel superalloys processed by Selective Laser Melting: A fundamental alloy design approach. *Acta Mater.* **2015**, *94*, 59–68. [CrossRef]
11. Tucho, W.M.; Cuvillier, P.; Sjolyst-Kverneland, A.; Hansen, V. Microstructure and hardness studies of Inconel 718 manufactured by selective laser melting before and after solution heat treatment. *Mater. Sci. Eng. A* **2017**, *689*, 220–232. [CrossRef]
12. Zhang, S.; Lin, X.; Wang, L.; Yu, X.; Hu, Y.; Yang, H.; Lei, L.; Huang, W. Strengthening mechanisms in selective laser-melted Inconel718 superalloy. *Mater. Sci. Eng. A* **2021**, *812*, 141145. [CrossRef]
13. Watring, D.S.; Benzing, J.T.; Hrabe, N.; Spear, A.D. Effects of laser-energy density and build orientation on the structure–property relationships in as-built Inconel 718 manufactured by laser powder bed fusion. *Addit. Manuf.* **2020**, *36*, 101425. [CrossRef]
14. Liu, S.Y.; Li, H.Q.; Qin, C.X.; Zong, R.; Fang, X.Y. The effect of energy density on texture and mechanical anisotropy in selective laser melted Inconel 718. *Mater. Des.* **2020**, *191*, 108642. [CrossRef]
15. Balbaa, M.; Mekhiel, S.; Elbestawi, M.; McIsaac, J. On selective laser melting of Inconel 718: Densification, surface roughness, and residual stresses. *Mater. Des.* **2020**, *193*, 108818. [CrossRef]
16. Yang, H.; Meng, L.; Luo, S.; Wang, Z. Microstructural evolution and mechanical performances of selective laser melting Inconel 718 from low to high laser power. *J. Alloys Compd.* **2020**, *828*, 154473. [CrossRef]
17. Du, D.; Dong, A.; Shu, D.; Zhu, G.; Sun, B.; Li, X.; Lavernia, E. Influence of build orientation on microstructure, mechanical and corrosion behavior of Inconel 718 processed by selective laser melting. *Mater. Sci. Eng. A* **2019**, *760*, 469–480. [CrossRef]
18. Teng, Q.; Li, S.; Wei, Q.; Shi, Y. Investigation on the influence of heat treatment on Inconel 718 fabricated by selective laser melting: Microstructure and high temperature tensile property. *J. Manuf. Process.* **2021**, *61*, 35–45. [CrossRef]
19. Xu, Z.; Cao, L.; Zhu, Q.; Guo, C.; Li, X.; Hu, X.; Yu, Z. Creep property of Inconel 718 superalloy produced by selective laser melting compared to forging. *Mater. Sci. Eng. A* **2020**, *794*, 139947. [CrossRef]
20. Gao, Y.; Zhang, D.; Cao, M.; Chen, R.; Feng, Z.; Poprawe, R.; Schleifenbaum, J.H.; Ziegler, S. Effect of δ phase on high temperature mechanical performances of Inconel 718 fabricated with SLM process. *Mater. Sci. Eng. A* **2019**, *767*, 138327. [CrossRef]
21. Zhang, S.; Wang, L.; Lin, X.; Yang, H.; Li, M.; Lei, L.; Huang, W. Precipitation behavior of δ phase and its effect on stress rupture properties of selective laser-melted Inconel 718 superalloy. *Compos. Part B Eng.* **2021**, *224*, 109202. [CrossRef]
22. Shi, J.J.; Li, X.; Zhang, Z.X.; Cao, G.H.; Russell, A.M.; Zhou, Z.J.; Li, C.P.; Chen, G.F. Study on the microstructure and creep behavior of Inconel 718 superalloy fabricated by selective laser melting. *Mater. Sci. Eng. A* **2019**, *765*, 138282. [CrossRef]
23. Zhang, D.; Niu, W.; Cao, X.; Liu, Z. Effect of standard heat treatment on the microstructure and mechanical properties of selective laser melting manufactured Inconel 718 superalloy. *Mater. Sci. Eng. A* **2015**, *644*, 32–40. [CrossRef]
24. *E139-11*; Standard Test Methods for Conducting Creep, Creep-Rupture, and Stress-Rupture Tests of Metallic Materials. ASTM International Publish: West Conshohocken, PA, USA, 2018.
25. Bean, G.E.; Witkin, D.B.; McLouth, T.D.; Patel, D.N.; Zaldivar, R.J. Effect of laser focus shift on surface quality and density of Inconel 718 parts produced via selective laser melting. *Addit. Manuf.* **2018**, *22*, 207–215. [CrossRef]
26. Song, K.; Yu, K.; Lin, X.; Chen, J.; Yang, H.; Huang, W. Microstructure and mechanical properties of heat treatment laser solid forming superalloy Inconel 718. *Jinshu Xuebao/Acta Metall. Sin.* **2015**, *51*, 935–942. [CrossRef]
27. Kumara, C.; Balachandramurthi, A.R.; Goel, S.; Hanning, F.; Moverare, J. Toward a better understanding of phase transformations in additive manufacturing of Alloy 718. *Materialia* **2020**, *13*, 100862. [CrossRef]
28. Zhou, K.; Wang, Z.; He, F.; Liu, S.; Li, J.; Kai, J.-j.; Wang, J. A precipitation-strengthened high-entropy alloy for additive manufacturing. *Addit. Manuf.* **2020**, *35*, 101410. [CrossRef]

29. Li, J.; Zhao, Z.; Bai, P.; Qu, H.; Liu, B.; Li, L.; Wu, L.; Guan, R.; Liu, H.; Guo, Z. Microstructural evolution and mechanical properties of IN718 alloy fabricated by selective laser melting following different heat treatments. *J. Alloys Compd.* **2019**, *772*, 861–870. [CrossRef]
30. Sui, S.; Tan, H.; Chen, J.; Zhong, C.; Li, Z.; Fan, W.; Gasser, A.; Huang, W. The influence of Laves phases on the room temperature tensile properties of Inconel 718 fabricated by powder feeding laser additive manufacturing. *Acta Mater.* **2019**, *164*, 413–427. [CrossRef]
31. Paul, S.; Liu, J.; Strayer, S.T.; Zhao, Y.; Sridar, S.; Klecka, M.A.; Xiong, W.; To, A.C. A Discrete Dendrite Dynamics Model for Epitaxial Columnar Grain Growth in Metal Additive Manufacturing with Application to Inconel. *Addit. Manuf.* **2020**, *36*, 101611. [CrossRef]
32. Yuan, C.; Lu, F.; Zhang, K.; Nie, P.; Hosseini, R.; Feng, K.; Li, Z. Dendritic microstructure and hot cracking of laser additive manufactured Inconel 718 under improved base cooling. *J. Alloys Compd.* **2016**, *670*, 312–321. [CrossRef]
33. Zhang, S.; Wang, L.; Lin, X.; Yang, H.; Huang, W. The formation and dissolution mechanisms of Laves phase in Inconel 718 fabricated by selective laser melting compared to directed energy deposition and cast. *Compos. Part B Eng.* **2022**, *239*, 109994. [CrossRef]
34. Zhang, S.; Lin, X.; Wang, L.; Yu, X.; Yang, H.; Lei, L.; Huang, W. Influence of grain inhomogeneity and precipitates on the stress rupture properties of Inconel 718 superalloy fabricated by selective laser melting. *Mater. Sci. Eng. A* **2021**, *803*, 140702. [CrossRef]
35. Cao, G.H.; Sun, T.Y.; Wang, C.H.; Li, X.; Liu, M.; Zhang, Z.X.; Hu, P.F.; Russell, A.M.; Schneider, R.; Gerthsen, D.; et al. Investigations of γ', γ'' and δ precipitates in heat-treated Inconel 718 alloy fabricated by selective laser melting. *Mater. Charact.* **2018**, *136*, 398–406. [CrossRef]

Disclaimer/Publisher's Note: The statements, opinions and data contained in all publications are solely those of the individual author(s) and contributor(s) and not of MDPI and/or the editor(s). MDPI and/or the editor(s) disclaim responsibility for any injury to people or property resulting from any ideas, methods, instructions or products referred to in the content.

coatings

MDPI

Article

Effect of Alternating Magnetic Field on the Organization and Corrosion Resistance of 2205 Duplex Stainless Steel Narrow-Gap Laser-MIG Hybrid Weld Head

Zhenxing He [1], Yong Zhao [1,*], Juan Fu [1], Fugang Chen [1], Guoqiang Chen [2] and Yonghui Qin [2]

1 Provincial Key Lab of Advanced Welding Technology, Jiangsu University of Science and Technology, Zhenjiang 212003, China; 211210601107@stu.just.edu.cn (Z.H.); fujuan@just.edu.cn (J.F.); chenfugang@just.edu.cn (F.C.)

2 Jiangsu Yangzi-Mitsui Shipbuilding Co., Ltd., Taicang 215400, China; chenguoqiang@yzjship.com (G.C.); qyh@yzjship.com (Y.Q.)

* Correspondence: yongzhao418@just.edu.cn; Tel.: +86-13815485895

Abstract: In this study, an alternating magnetic field is applied in the narrow-gap laser-MIG hybrid welding of 2205 duplex stainless steel with a thickness of 25 mm to achieve the purpose of balancing the ration of the two phases, refining the grains and improving the corrosion resistance. With the help of OM, EBSD, TEM, and other microstructural analysis methods, the organization evolution of a 2205 duplex stainless steel narrow-gap laser arc hybrid weld under the effect of alternating magnetic field is revealed. The corrosion resistance of the welded joints is investigated by electrochemical tests. The results show that the use of a 40 mT applied alternating magnetic field can not only effectively inhibit the generation of porosity and unfused defects in the weld, but also that the addition of an alternating magnetic field improves the ratio of austenite to ferrite in the weld, and the ratio of the two phases is increased from 0.657 without a magnetic field to 0.850. The weld grain preferential orientation is affected by the magnetic field, and the weld austenite grains are shifted from the Goss texture to the Copper texture. Under the electromagnetic stirring effect of the applied magnetic field, the average austenite grain size decreased from 4.15 μm to 3.82 μm, and the average ferrite grain size decreased from 4.99 μm to 4.08 μm. In addition, the effect of the alternating magnetic field increases the density of twins in the organization. Electrochemical test results show that the addition of an alternating magnetic field increases the corrosion potential by 75.2 mV and the pitting potential by 134.5 mV, which indicates that the corrosion resistance of the cover-welded specimens is improved by the effect of an alternating magnetic field. The improvement in corrosion resistance mainly depends on the austenite grain refinement and the increase in the austenite content.

Keywords: duplex stainless steel; alternating magnetic field; laser-MIG hybrid welding; microstructure evolution; corrosion

1. Introduction

Duplex stainless steel fully combines the advantages of austenitic and ferritic stainless steels due to the similar ratio of austenitic (γ) phase and ferrite (δ) in its microstructure [1,2]. As an iron-based alloy with both austenitic and ferritic organizations, it has both excellent mechanical properties and corrosion resistance. Therefore, it is widely used in production conditions such as oil storage and transportation, natural gas plants, and desalination plants [3,4]. Fusion welding is one of the commonly used traditional welding processes. However, due to the high heat input and fast cooling rate of the welding process, two-phase organization imbalance and joint property deterioration are prone to occur after the welding of duplex stainless steel [5,6]. High-energy density welding processes, such as narrow-gap laser-MIG hybrid welding, have received widespread attention due to the advantages of a small bevel size, less filler metal, a high welding efficiency, etc [7,8].

However, during the solidification process of the weld pool in narrow-gap laser-MIG hybrid welding, the laser energy mainly gathers at the bottom of the weld, and defects such as porosity and unfused welds often appear in the weld [9,10]. Therefore, it is necessary to adopt a practical and effective welding method to inhibit the generation of defects. Wang et al. [11] utilize a magnetic field to rotate the arc within a narrow-gap bevel, avoiding unfused defects through changes in energy distribution. Similarly, the magnetic field-assisted welding method also has an obvious inhibiting effect on porosity defects [12–15]. According to previous studies, the addition of a magnetic field to optimize the molten pool fluidity during the welding process is considered to be an effective way to improve the quality of welded joints [16–18]. Fu et al. [19] used laser arc hybrid welding of an S32101 duplex stainless steel plate under the magnetic field generated by Nd-Fe-B permanent magnets. The welding process is analyzed by high-speed imaging and spectral analysis. The results show that not only can the plasma temperature be increased under the effect of the magnetic field, but also that the melt pool flow is more uniform under a certain magnetic field strength. Applying the magnetic field assistance in the welding process and utilizing the electromagnetic stirring (EMS) effect generated by the magnetic field can achieve the purpose of refining the grain [20,21] and balancing the ratio of the two phases [22–24]. Zhu et al. [25] applied a perpendicular magnetic field in 316 L narrow-gap laser-MIG hybrid welding. The magnetic field ampere force was utilized to change the convection flow in the molten pool, causing the ferrite to remelt by dendrite fracture. In addition, the corrosion resistance of the metal is also very important [26]. Magnetic fields are used in the fusion welding of duplex stainless steel to improve the resistance to localized corrosion by using a low-intensity magnetic field assist [27]. Zhong et al. [28] utilized a novel in-situ transverse magnetic field acting on a cold metal transfer in a directional energy deposition arc of 316 L stainless steel. Under the action of the magnetic field, the Cr and Ni metal oxides in the passivation film increased significantly, and the stability of the passivation film was improved.

The addition of a magnetic field changes the distribution of the weld heat source, which affects the organization and properties [29]. Studies related to alternating magnetic fields are rare, and the use of magnetic field effects to improve organization and properties is valuable. In this study, 25 mm thick 2205 duplex stainless steel is selected as the base material, and an alternating magnetic field parallel to the weld is applied to narrow-gap laser-MIG hybrid welding. The effects of the alternating magnetic field on the two-phase equilibrium of the weld tissue, the distribution of the texture, the grain orientation, and the grain size are discussed via OM, EBSD, and TEM [30]. To elucidate the mechanism of the tissue structure transformation under the effect of alternating magnetic fields [31], we analyze the effect of an alternating magnetic field on the corrosion resistance of 2205 duplex stainless steel welded joints with narrow-gap laser arc hybrid welding through electrochemical tests. The effect of the alternating magnetic field on the organization characteristics and corrosion resistance of laser-MIG hybrid welded joints of duplex stainless steel is illustrated according to the test results. The roadmap for the paper is shown in Figure 1.

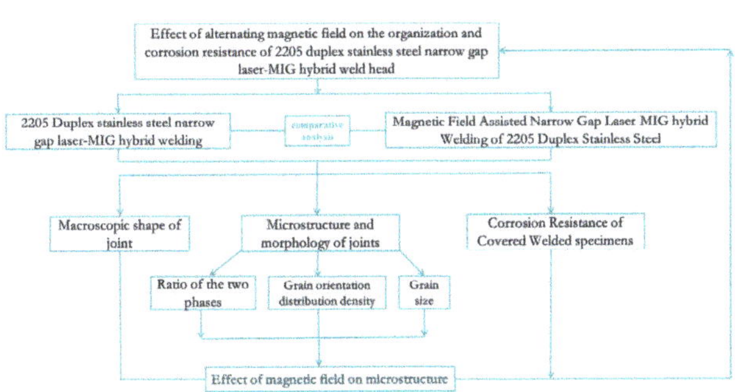

Figure 1. The roadmap for the paper.

2. Experimental Procedure

2.1. Base Material and Welding Procedure

In this study, a 25 mm thick 2205 duplex stainless steel plate is used as the base material, and the filler wire is selected as an ER2209 stainless steel solid wire with a diameter of 1.2 mm. The chemical composition of DSS2205 and ER2209 is shown in Table 1.

Table 1. Main chemical composition of DSS2205 and ER2209 (wt.%).

Material	C	S	P	Ni	Cr	Mo	N	Mn	Si	Fe
2205	0.024	0.001	0.023	5.68	22.39	3.13	0.17	1.38	0.39	Bal.
ER2209	0.019	0.0008	0.016	9.41	22.59	3.1	0.16	1.66	0.17	Bal.

In this study, the welding system consists of a YLS-6000 fiber laser generator (wavelength 1065–1080 nm), a Kuka robot, and a Fronius TPS500i PLUSE-type intelligent arc welding power system (Fronius welding Tech., Ltd., Zhuhai, China). Multi-position welding can be realized by a KUKA 6-axis robot. A fiber optic transmission system delivers the laser to the laser head and the welding is carried out using a KUKA industrial robot equipped with a laser head, wire feed mechanism, and protective gas unit. The alternating magnetic field generating equipment is an 8080 magnetic field controller from Jetline Engineering (Irvine, CA, USA) with a 4604 head. The alternating magnetic field assisted 25 mm thick 2205 duplex stainless steel narrow-gap laser-MIG hybrid welding equipment and groove size schematic are shown in Figure 2.

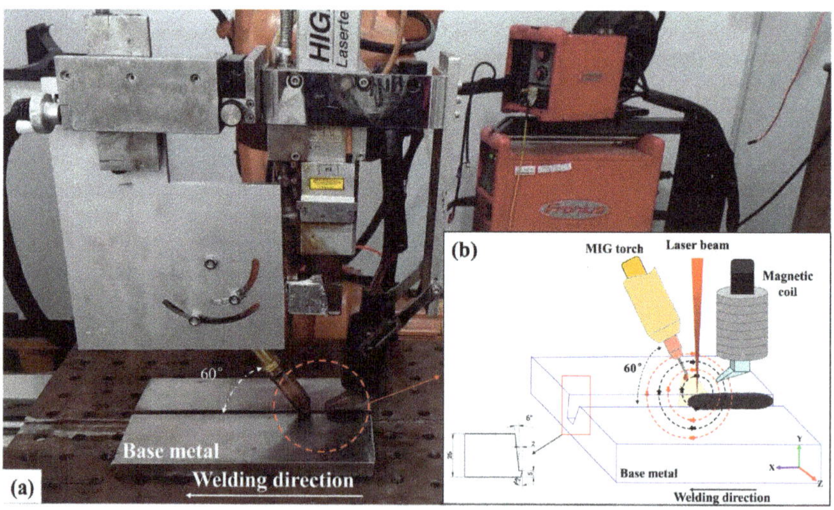

Figure 2. Schematic representation of the test setup and welding process: (**a**) Alternating magnetic field assisted narrow-gap laser arc hybrid welding test platform; (**b**) Schematic diagram of the test process and groove dimensions.

The size of the magnetic field strength and frequency can be changed by adjusting the magnetic field frequency and magnetic field amplitude knobs on the magnetic field controller, and by acting on the arc through the magnetic head, the regulation of the transition of the molten droplet and the flow of the molten pool can be realized. In the welding process, the magnetic field equipment can provide the highest alternating magnetic field frequency of 50 Hz, and the maximum alternating magnetic induction strength of 60 mT. The laser-MIG hybrid welding parameters are shown in Table 2. A magnetic field strength of 40 mT is chosen, which is measured by means of a Gauss meter. The magnetic field frequency is 15 Hz.

Table 2. Welding parameters of 25 mm DSS narrow-gap laser arc hybrid welding.

No.	P (kW)	I (A)	V (m/min)	Δf (mm)	D_{LA} (mm)	B (mT)	f (Hz)
1	2.3	220					
2	2.0	220					
3~4	2.0	240	0.8	+5	2	40	15
5~6	2.0	260					
7	1.7	240					

2.2. Microstructure Characterization

In this paper, the macroscopic morphology and microstructure of the welds are observed and analyzed by a 3D macroscope and metallurgical microscope, and Image Pro 6.0 image processing software is applied to statistically calculate the ratio of ferrite and austenite phases in different regions. The specimens are corroded with Beraha's reagent (0.5 g $K_2S_2O_5$ + 20 mL HCl + 80 mL of deionized water) after abrasive polishing and the duration of the corrosion is 5 to 10 s. The EBSD specimens are coarsely polished and then finely polished with a 50 nm SiO_2 suspension until there are no scratches under a 500× microscope, and vibratory polishing is used to remove the small amount of deformation and residual stresses left behind by the mechanical polishing. A Merlin Compact scanning electron microscope (Carl ZEISS Microscopy GmbH, Oberkochen, Germany) is used for the EBSD test with a test scan step of 1 μm, and the characteristics of the specimens' texture and grain orientation are comparatively analyzed by using Channel 5 software with and without the effect of an alternating magnetic field. A JEM-2100F field emission transmission electron microscope (TEM, JEOL, Tokyo, Japan) is used to analyze the two-phase deformation mechanism and interface morphology within the weld.

2.3. Corrosion Testing

Electrochemical tests are conducted using a three-electrode system, and the CS-2350H electrochemical workstation is used for electrochemical testing. A weld specimen of size 10 × 10 × 5 mm is prepared, and the copper conductor is fixed with the specimen using double-sided conductive copper foil tape and cold set with epoxy resin and then polished. The corrosion medium is artificial seawater (26.5 g/L NaCl, 24 g/L $MgCl_2$, 0.73 g/L KCl, 3.3 g/L $MgSO_4$, 0.2 g/L $NaHCO_3$, 1.1 g/L $CaCl_2$ and 0.28 g/L NaBr.). Before testing, to ensure the stability of the specimens in the medium, the specimens are placed in a simulated seawater medium for 0.5 h to ensure that there are no abnormal fluctuations in the open-circuit potential. The dynamic potential scanning range is −0.2 V~1.5 V, the rate is 0.333 mV/s, and the test temperature is 20 °C.

3. Results and Discussion

3.1. Weld Shaping and Microstructure Morphology

The shape of the narrow-gap laser-MIG hybrid weld seam assisted by an alternating magnetic field is shown in Figure 3. Under the effect of no magnetic field, the weld bead appears as a biting edge phenomenon, while the weld seam is smooth and beautiful in the case of an assisted magnetic field. Referring to the standard NB/T47013, X-ray nondestructive testing of the weld was performed. There are many pores and unfused defects inside the weld without a magnetic field. In contrast, there are no obvious defects inside the weld under a certain intensity and frequency. This is because the addition of the magnetic field produces an electromagnetic stirring effect on the molten pool, which increases the molten pool fluidity and is conducive to the overflow of porosity. At the same time, the magnetic field increases the plasma temperature, which effectively reduces the generation of unfused defects. This has been pointed out in the study of Cai et al. [32].

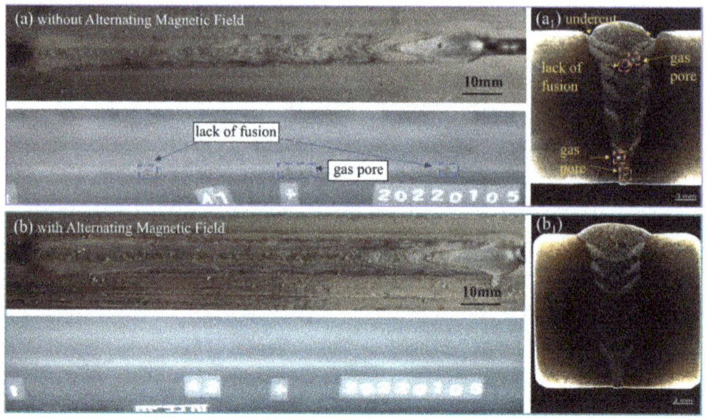

Figure 3. Welded joint forming and macroscopic morphology: (**a**,**a1**) without alternating magnetic field; (**b**,**b1**) with alternating magnetic field.

Narrow-gap laser arc hybrid welded joints for thick plates can be divided into four areas: the cosmetic bead, the filling bead, the overlapping bead, and the heat-affected zone. The organization and two-phase ratios of the four zones are analyzed. For the selected specimens, 10 collection locations are randomly selected to ensure that the collection area would cover most of the microstructure metallograph. The relative content of austenite and ferrite is measured using Image Pro Plus 6.0 software. The relative contents of austenite and ferrite are averaged over the 10 acquisition locations as shown in Table 3.

Table 3. Statistical table of two-phase organization score measurements.

Magnetic Field	Phase (%)	CB	FB	OB	HAZ
Without AMF	γ	36.5	40.1	46.1	35.7
	δ	63.5	59.9	53.9	64.3
With AMF	γ	39.9	43.4	53.6	40.2
	δ	60.1	56.6	46.4	59.8

In this paper, for illustrative purposes, the ratio of the relative austenite content to relative ferrite content is denoted as Y. The closer Y is to 1, the more balanced is the ratio of the two phases. The microstructure is shown in Figure 4.

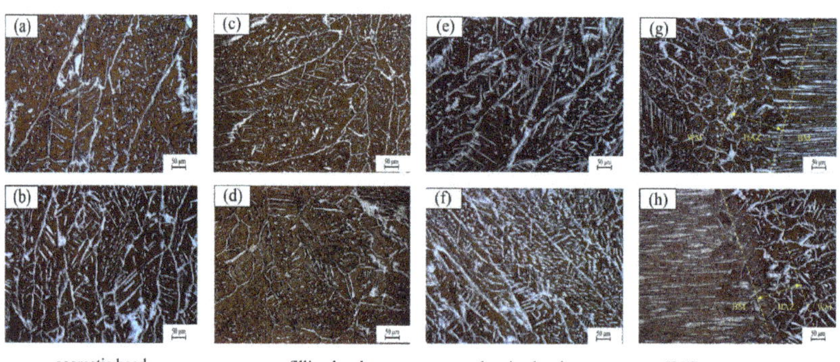

Figure 4. Microstructure and morphology: (**a**,**c**,**e**,**g**) without alternating magnetic field; (**b**,**d**,**f**,**h**) with alternating magnetic field.

Figure 4a shows the microstructure of the cosmetic bead without magnetic field assistance. Under the magnetic field, as in Figure 4b, the two-phase homogeneity is improved from 0.575 to 0.664. Higher homogeneity occurs in filler weld passes, as in Figure 4c,d, which is attributed to the combined effect of the magnetic field and preheating of the previous weld beads, which retards the solid-state phase transformation. The continuous nucleation of austenite increases the Y from 0.669 to 0.767 in the organization of the filling beads under a magnetic field. In multi-layer multi-pass welding, the overlap zone is affected by in-heating tempering, and there will be secondary austenite (γ_2) produced, which is distributed in the γ_1 primary austenite grain boundaries, with γ_1 expanding into the ferrite grains; γ_2 is finer in size, and the adjacent γ_2 shows a twin-crystal organization.

The austenite phase interface is very dense under the effect of an alternating magnetic field, and the organizational characteristics are dominated by the austenite phase. Figure 4f shows the organization and morphology under the alternating magnetic field, in which there is grain boundary austenite with continuous nucleation growth along the ferrite phase interface. There is densely present parallel lath-like Widmanstätten austenite, and dense intragranular austenite is also present within the grains. In addition, the stirring effect of the alternating magnetic field exacerbates the energy changes inside the molten pool, which promotes the precipitation of fine austenite within the ferrite grains in a non-diffusive transformation. Calculation of the two-phase homogeneity of austenite/ferrite in overlapping beads showed a two-phase homogeneity Y of 0.855 without the alternating magnetic field and Y of 1.160 with the applied magnetic field.

The heat-affected zone is not significantly affected by the magnetic field, but the homogeneity also increases from 0.555 to 0.672. The average two-phase homogeneity of the combined four zones improves from 0.663 to 0.816, which is closer to 1.0. This indicates that the effect of the alternating magnetic field leads to a more balanced ratio of the two phases after welding. Increased homogeneity of the two phases can improve the strength and plasticity of duplex stainless steel.

3.2. Distribution of Two-Phase Orientation and Texture in Welds

Figure 5 shows the EBSD phase diagram and the IPF of the cosmetic bead in the presence or absence of an alternating magnetic field. The IPF in the RD direction reflects the grain orientation distribution, and the EBSD phase diagram gives the distribution and content of the austenite and ferrite phases, where ferrite is red and austenite is yellow.

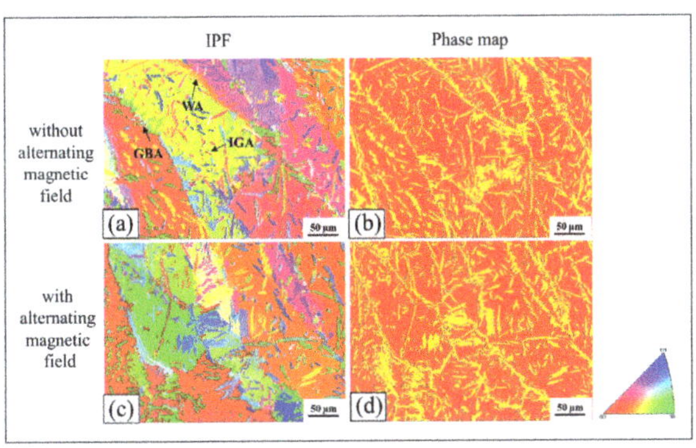

Figure 5. EBSD phase maps and IPFs in the RD direction of the cosmetic bead: (**a**,**b**) without alternating magnetic field; (**c**,**d**) with alternating magnetic field.

In the absence of an alternating magnetic field, the cosmetic bead is mainly a columnar crystal. It grows from the fusion line to the center of the weld. The weld organization mainly exhibits coarse ferrite grains and fine strips of GBA (grain boundary austenite) distributed at the grain boundaries. as shown in Figure 5a. From the grain boundary to the intracrystalline growth of the side plate stripes of WA (Widmanstätten austenite) and intracrystalline fine IGA (intragranular austenite), the overall diffuse distribution, when the ferrite grains are mainly oriented to (001)//RD. From the IPF of the weld with an applied alternating magnetic field in Figure 5c, it can be found that the weld is still dominated by coarse columnar crystals growing obliquely toward the center of the weld in general. However, the shape of the columnar crystals is irregular, and there is a transversely distributed austenite within the crystals, and the austenite is increased. The grain orientation is also gradually transformed to (101)//RD, and the grain orientation is (001)//RD and (101)//RD.

Comparison of the phase diagrams with and without the effect of an applied alternating magnetic field reveals that in the absence of a magnetic field, elongated GBA is distributed on the ferrite matrix and lath-like WA nucleates at the grain boundaries, and that it grows toward the grain interior. At the same time, IGA precipitates inside the ferrite as shown in Figure 5b. After the application of an alternating magnetic field, part of the region of the grain boundary characteristics became less obvious, the alternating magnetic field in the weld introduced dislocations or twins, which divided the original grain, so that the large-angle grain boundaries were split into a large number of small-angle grain boundaries and the ferrite grain was refined. The austenite phase in the weld grows, part of the austenite phase to the grain growth; there is a tendency to cross the ferrite grain, part of the austenite derived from the secondary austenite; the formation of the austenite phase is a feathery flocculent, as shown in Figure 5d.

Figure 6 shows the pole figure and the IPF of the austenite and ferrite {100}, {110}, and {111} within the weld with and without the action of an alternating magnetic field. The RD surface is usually chosen as the projection surface for the pole figure, and the pole figure is obtained by the pole density distribution. The distribution of grain surfaces on the rolling surface is analyzed by {100}, {110}, and {111} pole figures, and grain orientation is analyzed using the IPF in the RD direction.

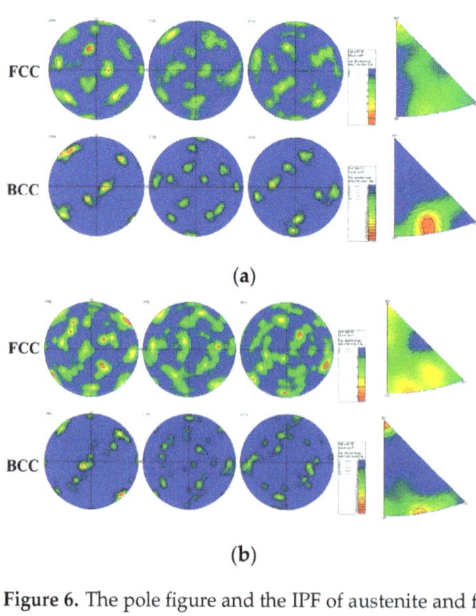

(**a**)

(**b**)

Figure 6. The pole figure and the IPF of austenite and ferrite. (**a**) without alternating magnetic field; (**b**) with alternating magnetic field.

From Figure 6a, it can be seen that in the absence of a magnetic field, there is a partial texture in the austenite, mainly with the {110} plane parallel to the rolling plane. The IPF shows that most of the grains are preferentially oriented in the <001> direction, with a maximum grain orientation distribution density of 5.88. Some dispersed texture also exists in the ferrite, and the main texture shows that the {100} plane is parallel to the rolling plane. The antipodal diagram shows that the preferential orientation of the grains is close to the

<101> direction, and the maximum grain orientation distribution density is 19.82. The pole figure and the IPF of the weld after the application of the alternating magnetic field are shown in Figure 6b, where the austenite weave density of the weld is enhanced under the action of the alternating magnetic field, and the austenite phase exhibits more selective orientations, close to <001>, <101> and gradually shifting to <111> in the direction along the rolling direction. The austenite grains may be shifted from Goss texture to Copper texture, and the grain orientation distribution density increased to 6.46.

After the application of an alternating magnetic field, there is no significant change in the main ferrite texture. However, as shown in the IPF, the grain orientation shifted from <101> to <001>, at which time the grain orientation distribution density increased to 21.60.

3.3. *Grain Boundary Orientation Difference and Grain Size of the Weld*

According to the difference in the angle of the orientation difference between neighboring grains, grain boundaries can be classified into low angle grain boundaries, LABs (orientation difference <15°) and high angle grain boundaries, HABs (orientation difference >15°), of which HAGBs include normal large angle grain boundaries and coincident site lattice grain boundaries; generally, in austenite often appear [111]/60° of Σ3 coincident lattice [33,34], and the orientation relationship of the two CSLs is twinning, so in this paper, the CSL grain boundaries content will be recognized as Σ3 twinning boundary content.

Figure 7 shows the distribution of the grain boundary orientation difference between the two phases of the cosmetic bead with and without the effect of an alternating magnetic field. Without the effect of alternating magnetic fields, the ferrite phase contains 39.4% LAGBs, 60.6% HAGBs, and the twinned grain boundary is 1.82%. The austenite phase contains 52.7% HAGBs and 3.0% twin boundaries, at which time the grain orientation is easily changed in favor of the Goss texture. After the application of an alternating magnetic field, the HAGBs in the ferrite phase decrease to 44.7%, and the HAGBs in the austenitic phase increase. The effect of the alternating magnetic field impacts the dislocation climbing within the ferrite and reduces the laminar dislocation energy. The action on the austenitic phase favors the formation of twins, increases the grain boundary transition potential, and raises the possibility of a grain orientation change. As a result of the analysis of the contrast pole figure, the austenite grains are transferred from the Goss texture to the Copper texture. The grain size distribution of the two phases of the weld with and without the assistance of the alternating magnetic field is shown in Figure 8. The average grain size of the ferrite phase of the weld without an alternating magnetic field is 4.99 μm, and the average grain size of the austenite phase is 4.15 μm. After the application of the alternating magnetic field, the ferrite phase and the austenite phase are affected by electromagnetic stirring, and the average sizes of the grains of the two phases become smaller, with the ferrite phase decreasing to 4.08 μm, and the austenite phase decreasing to 3.82 μm. It is worth mentioning that the grain refinement increases the weld toughness. The refinement of austenite and the increase in austenite content in the cover organization play a very important role in the improvement of corrosion resistance.

Figure 9 shows the TEM and SEAD with and without a magnetic field. Figure 9a shows the TEM photographs of the weld with no alternating magnetic field, and the twinned organization with elongated lamellar structure throughout the grain is observed within the austenitic organization. Figure 9b shows the TEM pictures of the weld after the introduction of the magnetic field. After the introduction of an alternating magnetic field, the twin density is increased to some extent, which leads to a texture transition when the density reaches a certain level. As a result, twin crystals are more easily formed with the assistance of the alternating magnetic field, and the possibility of grain orientation change is increased.

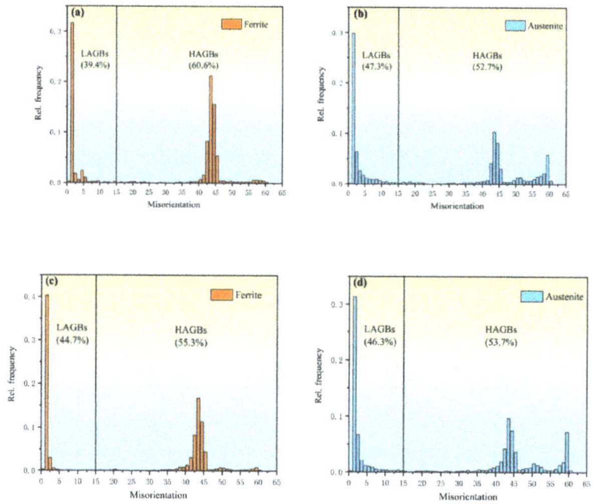

Figure 7. Misorientation angle distribution of grain boundaries of the cosmetic bead: (**a**) Ferrite without alternating magnetic field; (**b**) Austenite without alternating magnetic field; (**c**) Ferrite without alternating magnetic field; (**d**) Austenite without alternating magnetic field.

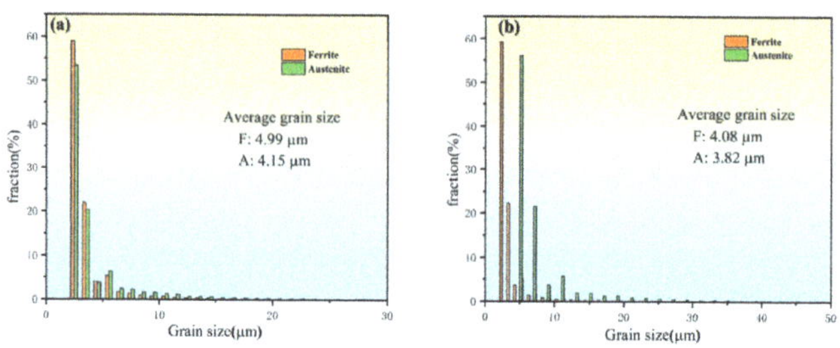

Figure 8. Grain size of the two phases of the weld: (**a**) without alternating magnetic field; (**b**) with alternating magnetic field.

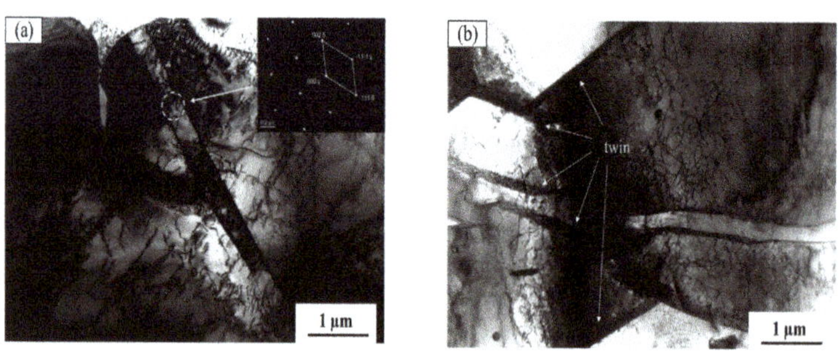

Figure 9. Twin crystal morphology in the weld: (**a**) without alternating magnetic field; (**b**) with alternating magnetic field.

3.4. Potentiodynamic Polarization

Dynamic potential scanning tests are carried out on welds with and without magnetic fields to study the corrosion resistance of the welds. The polarization curves are shown in Figure 10. In the dynamic potential polarization curve test, the corrosion potential (E_{corr}) and corrosion current density (I_{corr}) are commonly used to indicate the dissolution ability of the metal, which visually reflects the corrosion resistance of the metal under no applied conditions. Theoretically, the higher the value of E_{corr} is, the lower is the value of I_{corr}, indicating that the metal corrosion resistance is better [35]. Pitting corrosion of stainless steel requires special attention and generally occurs at inclusions or grain boundaries [36]. The anodic current undergoes an abrupt change when pitting occurs, and the potential inflection point obtained at this time is the pitting potential (E_b). The higher the pitting potential is, the better is the corrosion resistance of the metal.

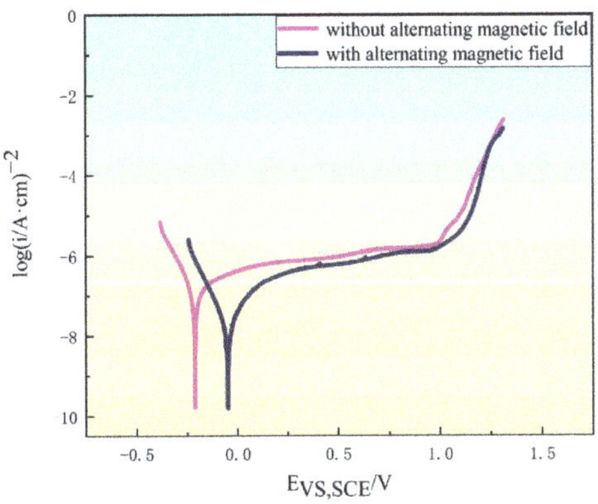

Figure 10. Polarization curve of welded joint in simulated seawater.

Figure 10 shows that the trends of the curves are similar for the welds with and without magnetic field assistance. With the increase in corrosion potential, both specimens experience long passivation intervals and then tend to stabilize. As the corrosion potential continues to rise, the current density of the welded specimen without the effect of an alternating magnetic field is the first to change and enter the over-passivation zone, and pitting corrosion occurs on the metal surface. Comparing the current density at the same corrosion potential, it can be found that the current density of the welded specimen with the applied alternating magnetic field is significantly lower than that without the alternating magnetic field, and the lower corrosion current indicates a slower corrosion rate and better corrosion resistance. Compared with the welded specimens without an alternating magnetic field, the welded specimens with an alternating magnetic field have higher corrosion potential; the pitting potential is obviously larger, which indicates that the 2205 duplex stainless steel welded joints with an alternating magnetic field have better corrosion resistance. The eigenvalue parameters of the dynamic potential polarization curves are obtained by fitting using the Tafel extrapolation method as shown in Table 4.

Table 4. The polarization curve fitting results and corrosion rate of cover welded specimens.

Magnetic Field	E_{corr} (V)	I_{corr} (A·cm^{-2})	E_{pass} (V)	I_{pass} (A·cm^{-2})	E_{pit} (V)	Corrosion Rate (mm/y)
Without AMF	−0.2150	1.82×10^{-7}	0.1439	6.26×10^{-7}	1.1775	2.14×10^{-3}
With AMF	−0.1398	2.64×10^{-8}	0.0637	4.21×10^{-7}	1.312	0.28×10^{-3}

In Table 4, the I_{corr} of the welded specimen with an alternating magnetic field is 2.64×10^{-8} A cm^{-2}, which is much lower than that of $I_{corr} = 1.82 \times 10^{-7}$ A cm^{-2} without an applied alternating magnetic field, and the E_{pass} is also significantly reduced after the magnetic field, which indicates that it is easier to form a stable passivation film. The increase in pitting potential indicates that the pitting resistance of the specimen is improved, and according to the I_{corr} calculation, the corrosion rate of the weld without magnetic field assistance is reduced from 0.00214 mm per year to 0.00028 mm per year. The above data indicate that the auxiliary effect of the alternating magnetic field not only makes it easier to form a stable passivation film but also significantly reduces the corrosion sensitivity of the welded specimen and improves the pitting resistance of the weld. The improvement in corrosion resistance is mainly due to the introduction of an alternating magnetic field. The electromagnetic stirring action increases the proportion of austenite in the weld.

The relevance of the research for practical implementation:

The use of magnetic field-assisted duplex stainless steel narrow-gap laser arc hybrid welding can not only refine the grain but also improve the uniformity of the organization, which is very important to solve the real corrosion problems. In addition, the magnetic field is generated by a simple and portable magnetic field generator, which can be adapted to a variety of welding environments. It is useful for the generalization of this study in engineering.

4. Conclusions

This paper mainly investigates the effect of alternating magnetic fields on the organization and corrosion resistance of a narrow-gap laser arc hybrid weld of duplex stainless steel and analyzes the weld microstructure, fabrication distribution, grain size, etc. The main conclusions are as follows:

1. Welded joints without an applied magnetic field suffer from unfused and porosity defects and poor weld shaping. The weld is well formed and free of unfused and porosity defects with the assistance of a 40 mT alternating magnetic field. This is due to the fact that the presence of the magnetic field changes the heat distribution of the laser arc hybrid heat source in the narrow-gap space.
2. The two-phase homogeneity obtained by combining the two-phase contents of the four regions of the weld is 0.850, which is closer to 1.0 than the two-phase homogeneity of 0.657 in the absence of an alternating magnetic field. This indicates that the overall ratio of the two phases in the weld with the alternating magnetic field is more balanced than that without the magnetic field.
3. Under the effect of alternating magnetic fields, the austenite grain orientation distribution density is enhanced from 5.88 to 6.46 and the ferrite grain orientation distribution density is enhanced from 19.82 to 21.69. The texture density of austenite is increased, while the texture density of ferrite does not change significantly.
4. The ferrite and austenite phases are affected by the electromagnetic stirring effect of the applied magnetic field, and the average grain size of the austenite phase is reduced from 4.15 μm to 3.82 μm. The average grain size of the ferrite phase is reduced from 4.99 μm to 4.08 μm. The addition of the magnetic field results in the grain refinement of both phases. In addition, the density of twins in the tissue increases under the effect of an alternating magnetic field.
5. The electrochemical test results show that the corrosion resistance and passivation ability of the weld are improved under the magnetic field-assisted action. The Ecorr of the weld under the magnetic field-assisted condition is 75.2 mV higher than that in the absence of a magnetic field, and the Epit is 134.5 mV higher than that in the absence of a magnetic field. The corrosion rate is about 1/7th of that without the magnetic field. The increase in corrosion resistance depends mainly on austenite changes including the austenite grain size and content.
6. The quality of the weld is good with no unfused or porosity defects. Under the action of the magnetic field, the two phases are balanced, the organization is refined, and the corrosion resistance is improved. This provides support for the application of magnetic fields in the field of duplex stainless steel.

Author Contributions: Z.H.: Conceptualization, Investigation, Data curation, Experiments, Writing—original draft preparation. Y.Z.: Investigation, Supervision, Funding acquisition, Writing—review and editing. J.F.: Data curation, Supervision, Writing—review and editing. F.C.: Software, Experiments. G.C.: Supervision, Funding acquisition. Y.Q.: Investigation, Funding acquisition. All authors have read and agreed to the published version of the manuscript.

Funding: This research was funded by Jiangsu Postgraduate Research and Practice Innovation Program Project of China by SJCX23_2177.

Institutional Review Board Statement: Not applicable.

Informed Consent Statement: Written informed consent has been obtained from the co-authors to publish this paper.

Data Availability Statement: The data used to support the findings of this study are available from the corresponding author upon request.

Acknowledgments: The authors would like to thank the Jiangsu Postgraduate Research and Practice Innovation Program Project of China (SJCX23_2177) for the financial support of this paper.

Conflicts of Interest: The authors declare that they have no known competing financial interests or personal relationships that could have appeared to influence the work reported in this paper.

References

1. Rosado-Carrasco, J.; Krupp, U.; López-Morelos, V.H.; Giertler, A.; García-Rentería, M.A.; González-Sánchez, J. Effect of a Magnetic Field Applied during Fusion Welding on the Fatigue Damage of 2205 Duplex Stainless Steel Joints. *Int. J. Fatigue* **2019**, *121*, 243–251. [CrossRef]
2. Wu, J.; Cao, F.J.; Sun, T.; Huang, G.Q.; Li, M.S.; Hou, W.T.; Piao, Z.Y.; Shen, Z.K.; Shen, Y.F. Developing Ultrafine-Grained Structure with Balanced α/γ Fraction via Underwater Friction Stir Processing Enables Enhanced Wear and Corrosion Resistance of Duplex Stainless Steel. *Surf. Coat. Technol.* **2023**, *457*, 129295. [CrossRef]
3. Chaudhari, A.N.; Dixit, K.; Bhatia, G.S.; Singh, B.; Singhal, P.; Saxena, K.K. Welding Behaviour of Duplex Stainless Steel AISI 2205: A Review. *Mater. Today Proc.* **2019**, *18*, 2731–2737. [CrossRef]
4. Luo, J.; Yan, P.; Fan, Y.; Luo, S.; Long, Y. Investigation of Corrosion Behavior of 2205 Duplex Stainless Steel Coiled Tubing in Complex Operation Environments of Oil and Gas Wells. *Eng. Fail. Anal.* **2023**, *151*, 107355. [CrossRef]
5. Li, F.; Liu, Y.; Ke, W.; Jin, P.; Kong, H.; Chen, M.; Sun, Q. A Novel Pathway to Weld Forming Control and Microstructure Improvement of Duplex Stainless Steel via Alternating Magnetic Field. *J. Manuf. Process.* **2022**, *80*, 581–590. [CrossRef]
6. Wu, M.; Xin, R.; Wang, Y.; Zhou, Y.; Wang, K.; Liu, Q. Microstructure, Texture and Mechanical Properties of Commercial High-Purity Thick Titanium Plates Jointed by Electron Beam Welding. *Mater. Sci. Eng. A* **2016**, *677*, 50–57. [CrossRef]
7. Ning, J.; Zhang, L.-J.; Yang, J.; Yin, X.-Q.; Wang, X.-W.; Wu, J. Characteristics of Multi-Pass Narrow-Gap Laser Welding of D406A Ultra-High Strength Steel. *J. Mater. Process. Technol.* **2019**, *270*, 168–181. [CrossRef]
8. Sokolov, M.; Salminen, A.; Kuznetsov, M.; Tsibulskiy, I. Laser Welding and Weld Hardness Analysis of Thick Section S355 Structural Steel. *Mater. Des.* **2011**, *32*, 5127–5131. [CrossRef]
9. Cho, W.-I.; Woizeschke, P. Analysis of Molten Pool Behavior with Buttonhole Formation in Laser Keyhole Welding of Sheet Metal. *Int. J. Heat Mass Transf.* **2020**, *152*, 119528. [CrossRef]
10. Zhao, Y.; Wang, W.; Yan, K.; Liu, C.; Zou, J. Microstructure and Properties of Cu/Ti Laser Welded Joints. *J. Mater. Process. Technol.* **2018**, *257*, 244–249. [CrossRef]
11. Wang, J.; Sun, Q.; Zhang, T.; Zhang, S.; Liu, Y.; Feng, J. Arc Characteristics in Alternating Magnetic Field Assisted Narrow Gap Pulsed GTAW. *J. Mater. Process. Technol.* **2018**, *254*, 254–264. [CrossRef]
12. Raoelison, R.N.; Racine, D.; Zhang, Z.; Buiron, N.; Marceau, D.; Rachik, M. Magnetic Pulse Welding: Interface of Al/Cu Joint and Investigation of Intermetallic Formation Effect on the Weld Features. *J. Manuf. Process.* **2014**, *16*, 427–434. [CrossRef]
13. Rong, Y.; Huang, Y.; Xu, J.; Zheng, H.; Zhang, G. Numerical Simulation and Experiment Analysis of Angular Distortion and Residual Stress in Hybrid Laser-Magnetic Welding. *J. Mater. Process. Technol.* **2017**, *245*, 270–277. [CrossRef]
14. Yin, X.; Gou, J.; Zhang, J.; Sun, J. Numerical Study of Arc Plasmas and Weld Pools for GTAW with Applied Axial Magnetic Fields. *J. Phys. Appl. Phys.* **2012**, *45*, 285203. [CrossRef]
15. Li, L.; Huang, C.; Han, G.; Chen, R. Recent Progress on External Magnetic Field Assisted Laser Welding: Mechanism, Effect and Technology. *Int. J. Adv. Manuf. Technol.* **2023**, *125*, 1–23. [CrossRef]
16. Yang, C.; Yang, F.; Meng, X.; Putra, S.N.; Bachmann, M.; Rethmeier, M. Phase-Field Simulation of the Dendrite Growth in Aluminum Alloy AA5754 during Alternating Current Electromagnetic Stirring Laser Beam Welding. *Int. J. Heat Mass Transf.* **2024**, *218*, 124754. [CrossRef]
17. Yue, J.; Zhou, H.; Xie, C.; Zhou, W.; Liu, W. Temperature Field Simulation and Asymmetric Heat Transfer Distribution of Dissimilar Steel Welded with External Transverse Magnetic Field. *Mater. Today Commun.* **2023**, *37*, 107141. [CrossRef]
18. Tan, C.; Liu, Y.; Xu, B.; Wang, H.; Liu, F.; Gong, X.; Zeng, Z.; Chen, B.; Song, X. Numerical and Experimental Study of Thermal Fluid Flow and Keyhole Dynamic in Laser Welding of Aluminum Alloy Assisted by Electromagnetic Field. *Opt. Laser Technol.* **2023**, *157*, 108718. [CrossRef]
19. Fu, J.; Zhao, Y.; Zou, J.; Liu, X.; Pan, Y. Influence of the Magnetic Field on the Melting and Solidification Behavior of Narrow-Gap Laser Welding with Filler Wire. *Int. J. Adv. Manuf. Technol.* **2022**, *123*, 1123–1131. [CrossRef]

20. Ding, L.; Wang, H.; Zeng, Z.; Vladyslav, K.; Chen, L.; Li, F.; Zhang, Y.; Qin, B. The Influence Mechanism of Alternating Magnetic Field on Microstructure and Mechanical Properties of Ti-6Al-4V Alloy. *J. Mater. Eng. Perform.* **2023**. [CrossRef]
21. Xu, T.; Shi, Y.; Jiang, Z.; Liu, B.; Zhan, J.; Wang, Z. An Extraordinary Improvement in Cryogenic Toughness of K-TIG Welded 9Ni Steel Joint Assisted by Alternating Axial Magnetic Field. *J. Mater. Res. Technol.* **2023**, *25*, 3071–3077. [CrossRef]
22. García Rentería, M.A.; López Morelos, V.H.; García Hernández, R.; Curiel López, F.; Lemus-Ruíz, J. Effect on the Microstructure and Mechanical Properties of the Electromagnetic Stirring during GMA Welding of 2205 DSS Plates. *Mater. Sci. Forum* **2013**, *755*, 61–68. [CrossRef]
23. Li, F.; Liu, Y.; Jin, P.; Chen, M.; Ke, W.; Du, Y.; Liu, J.; Sun, Q. Appreciable-Tuned Ferrite/Austenite Phase Balance in the Fusion Zone of GTAW Welds via an Assisted Magnetic Field. *J. Alloys Compd.* **2021**, *885*, 160851. [CrossRef]
24. Liu, L.; Zhu, Y.; Liu, R. Study of Twin Tungsten Electrode—Wire Electrode Indirect Arc Welding Assisted by Alternating Magnetic Field. *J. Manuf. Process.* **2023**, *101*, 171–180. [CrossRef]
25. Zhu, Z.; Ma, X.; Wang, C.; Mi, G. Grain Refinement and Orientation Alternation of 10 mm 316L Welds Prepared by Magnetic Field Assisted Narrow Gap Laser-MIG Hybrid Welding. *Mater. Charact.* **2020**, *164*, 110311. [CrossRef]
26. Thakur, A.; Kumar, A.; Sharma, S.; Ganjoo, R.; Assad, H. Computational and Experimental Studies on the Efficiency of Sonchus Arvensis as Green Corrosion Inhibitor for Mild Steel in 0.5 M HCl Solution. *Mater. Today Proc.* **2022**, *66*, 609–621. [CrossRef]
27. García-Rentería, M.A.; López-Morelos, V.H.; González-Sánchez, J.; García-Hernández, R.; Dzib-Pérez, L.; Curiel-López, F.F. Effect of Electromagnetic Interaction during Fusion Welding of AISI 2205 Duplex Stainless Steel on the Corrosion Resistance. *Appl. Surf. Sci.* **2017**, *396*, 1187–1200. [CrossRef]
28. Zhong, Y.; Zheng, Z.; Li, J.; Wang, C. Effect of In-Situ Transverse Magnetic Field on Microstructure, Mechanical Properties and Corrosion Resistance of the Directed Energy Deposited 316L Stainless Steel. *Addit. Manuf.* **2023**, *68*, 103508. [CrossRef]
29. Wang, L.; Ma, Y.; Xu, J. Numerical Simulation of Arc-Droplet-Weld Pool Behaviors during the External Magnetic Field-Assisted MIG Welding-Brazing of Aluminum to Steel. *Int. J. Therm. Sci.* **2023**, *194*, 108530. [CrossRef]
30. Zhu, Z.; Ma, X.; Mi, G.; Wang, C. Electron Microscopy Study of Laser Welded GH909 Superalloy Joint. *J. Mater. Res. Technol.* **2020**, *9*, 15525–15536. [CrossRef]
31. Santos, R.M.; Rodrigues, D.G.; Dias Santos, M.L.; Santos, D.B. Martensite Reversion and Strain Hardening of a 2304 Lean Duplex Stainless Steel Subjected to Cold Rolling and Isochronous Annealing at Low Temperatures. *J. Mater. Res. Technol.* **2022**, *16*, 168–186. [CrossRef]
32. Cai, B.; Fu, J.; Zhao, Y.; Chen, F.; Qin, Y.; Song, S. Effect of Alternating Magnetic Field on Arc Plasma Characteristics and Droplet Transfer during Narrow Gap Laser-MIG Hybrid Welding. *Metals* **2021**, *11*, 1712. [CrossRef]
33. Fu, J.; Rao, Z.; Zhao, Y.; Zou, J.; Liu, X.; Pan, Y. Microstructure and Texture Characterization of Duplex Stainless Steel Joints Welded by Alternating Magnetic Field-Assisted Hybrid Laser-GMAW Welding. *Materials* **2022**, *15*, 8741. [CrossRef]
34. Cui, S.; Pang, S.; Pang, D.; Zhang, Z. Influence of Welding Speeds on the Morphology, Mechanical Properties, and Microstructure of 2205 DSS Welded Joint by K-TIG Welding. *Materials* **2021**, *14*, 3426. [CrossRef]
35. You, J.; Li, Z.; Li, Y.; Cai, Z. Microstructure and Properties of Underwater Wet Laser Welded 0Cr25Ni6Mo3N Duplex Stainless Steel Joints. *Mater. Today Commun.* **2023**, *36*, 106622. [CrossRef]
36. Sun, Y.; Tan, X.; Lan, R.; Ran, G.; Li, J.; Jiang, Y. Mechanisms of Inclusion-Induced Pitting of Stainless Steels: A Review. *J. Mater. Sci. Technol.* **2024**, *168*, 143–156. [CrossRef]

Disclaimer/Publisher's Note: The statements, opinions and data contained in all publications are solely those of the individual author(s) and contributor(s) and not of MDPI and/or the editor(s). MDPI and/or the editor(s) disclaim responsibility for any injury to people or property resulting from any ideas, methods, instructions or products referred to in the content.

coatings

MDPI

Communication

Refining Micron-Sized Grains to Nanoscale in Ni-Co Based Superalloy by Quasistatical Compressive Deformation at High Temperature

Hui Xu [1,2,*], Yanjun Guo [1], Jiangwei Wang [2], Ze Li [1], Lu Wang [1], Xiaohui Li [1] and Zhefeng Zhang [3,*]

1 Materials Research Center, Huadian Electric Power Research Institute Co., Ltd., Hangzhou 310030, China
2 School of Materials Science and Engineering, Zhejiang University, Hangzhou 310027, China
3 Shi-Changxu Innovation Center for Advanced Materials, Institute of Metal Research, Chinese Academy of Sciences, Shenyang 110016, China
* Correspondence: hui-xu@chder.com (H.X.); zhfzhang@imr.ac.cn (Z.Z.)

Abstract: Compressive deformation was carried out in an Ni-Co-based superalloy with relatively low stacking fault energy (SFE) at 725 °C and a strain rate of 10^{-2} s^{-1}; the underlying micromechanisms were investigated under true compression strains varying from 0.1 to 1.0. It was found that dislocation slipping accompanied by stacking fault (SF) shearing dominated the compressive deformation under the strain of 0.1 and 0.2. As the strain increased to 0.3 and 0.4, microtwinning was activated and then interacted with dislocations, leading to the formation of dislocation tangles or blocky distorted region. When true strain was further increased to 0.6, abundant subgains (SGs) with polygonous shape appeared and then transformed into nanograins as true strain increased to 1.0. It is demonstrated that high strain and microtwinning are the prerequisites for the evolution of nanograins in the deformed Ni-Co-based superalloy. High strain can produce plentiful dislocations and distorted micro-sized SGs; then the microtwins sheared these distorted regions and refined the micro-sized SGs into nanoscale, which subsequently transformed into nanograins with further deformation.

Keywords: superalloy; nanograin; microtwin; dislocation; compression

1. Introduction

Ni-Co-based superalloys have been widely used in the discs of industrial gas turbines for their superior combination of service performance that includes excellent high-temperature strength, oxidation resistance, creep resistance, and fatigue properties [1]. In view of the unmatched mechanical properties of the present superalloys employed in gas turbines that demand a higher thrust–weight ratio and thermal efficiency, developing new superalloys with better service properties and designing advanced coatings with special protective effects are often seen as feasible approaches [2–4]. In recent years, a novel designing principle that strengthens superalloy at service temperatures and weakens superalloy at processing temperatures has been developed and applied in the modified Ni-Co-based superalloy by controlling its alloying element and stacking fault energy (SFE) [5–8]. SFE has a strong influence on the strengthening mechanisms, mechanical behaviors, and microstructural evolution of superalloys. Abundant studies have been carried out on the deformation behaviors of conventional superalloys with higher SFE during tensile or compressive tests. A consensus has been reached that dislocation motion dominates the deformation process at different conditions, and the precipitation strengthening of the γ' phase is greatly affected by the mode of dislocation slip [9,10]. The modified Ni-Co-based superalloys possess low SFE due to their high electron hole concentration adding Co element [11], which exhibits distinct deformation behaviors with respect to the high-SFE superalloys. In the low-SFE superalloys, it was proved that microtwins (MTs) could be introduced during tensile deformation [12]. Furthermore, the ultimate tensile strength and uniform elongation were improved synchronously because microtwinning was activated in superalloys with decreased SFE at 650 and 725 °C [13]. When deformation occurred at higher strains, nanograins (NGs) were found to be produced in the superalloys deformed at a high temperature and low strain rate, which resulted in higher flow stress [14]. However, the underlying mechanisms on the NG formation related to strain variation have not been clarified in these low-SFE superalloys based on the experimental analysis.

Moreover, grain refinement generally benefits the mechanical properties of bulk metal via strengthening and toughening the materials based on the Hall–Petch effect. In high-SFE metallic materials such as pure Ni, pure Fe, Cu, and Cu alloys, deformation at a low temperature and high strain rate can produce numerous NGs due to dislocation-mediated grain refining [15–18]. However, the nanocrystallization of modified Ni-Co-based superalloys shows them to be entirely different from these materials either in the deformation mechanisms or deformation conditions. For better understanding the underlying mechanisms of nanocrystallization with stain variation in Ni-Co-based superalloys, interrupted compressive tests were carried out for a newly developed Ni-Co-based superalloy in order to clarify the critical procedures for its grain refinement. NGs are beneficial for strengthening the Ni-Co-based superalloys, which can be utilized in the gradient structure of the superalloys that require surface strengthening.

2. Experimental

One kind of Ni-Co-based superalloy was selected as experimental materials with the chemical compositions of 14.6Cr-3.7Mo-20.5Co-1.9Al-5.7Ti-0.26Fe-0.03C-0.051Zr-Ni bal (wt.%). The master alloy ingots were first smelted in a 20 Kg vacuum arc furnace, and then homogenized heat treatment was performed to alleviate composition segregation. Then, round rods were cut from the treated ingots, which were subsequently encased and extruded into test bars with diameter of 35 mm at 1160 °C. Finally, the test bars were subjected to two-step solution treatments and two-step aging treatments for 1170 °C/4 h/AC + 1080 °C/4 h/AC + 845 °C/24 h/AC + 760 °C/16 h/AC (AC is air cooling), resulting in an average grain size of 124 μm determined by electron back-scattered diffraction analysis. Cylindrical specimens with dimensions of Φ 5 × 8 mm were machined by electrical discharge machining, followed by mechanical polishing to eliminate surface scratch. Compression tests were then carried out for the specimens at 725 °C and under the strain rate of 10^{-2} s^{-1} using a Gleeble 3800 thermal simulation test machine with temperature control accuracy of ±1 °C, displacement measurement sensitivity of ±1%, and force measurement accuracy of ±1%, which were interrupted at different true strains (ε) varying from 0.1 to 1.0 in order to study the dependence of deformation microstructures on compression strains. The microstructure was characterized by an FEI Tecnai F20 transmission electron microscope (TEM) operated at 200 kV. TEM slices with thickness of 500 μm were cut from the middle part of the compressed samples along cross sections. The slices were further ground down to 50 μm and perforated by a twin-jet electro-polisher in a solution of 10% perchloric acid and 90% ethanol under conditions of 30~32 V and −22~−20 °C.

3. Results

The original microstructures were composed of coarse primary γ' precipitates, tiny secondary γ' precipitates, intergranular carbides, and γ matrix (Figure 1). Two kinds of γ' precipitate were coherently embedded in the face-centered cubic γ matrix. As compressive strain increased, phase constituents showed no apparent changes, while the deformed microstructures evolved continually. The deformation microstructures of the tested sample at lower strains are depicted in Figure 2. It could be found that deformation was dominated by dislocation motion since high density of dislocation tangles was introduced at the strain of 0.1 (Figure 2a); meanwhile, fine SF debris scattered both in the γ matrix and primary γ' precipitates. It could be found that SF propagation was inhibited by the phase interface of the γ' precipitates, leading to the independent growth of SFs, which were restrained either in the γ matrix or primary γ' precipitates (Figure 2b). As the strain increased to 0.2, dislocation slip became more prevalent in order to accommodate larger compression strain; meanwhile, dislocation climb could be activated thermally when the slip of dislocations was hindered by the γ' precipitates. Dislocation motion that included slip and climb continued to dominate the compressive deformation since high density of dislocation tangles also appeared in the γ matrix (Figure 2c). The nucleation and propagation of SFs seemed to be enhanced in that multidirectional growth of SFs could be detected both in the γ matrix and primary γ' precipitates (Figure 2d). Meanwhile, the length of SFs generated from different slip systems increased obviously, and the SFs sheared with each other when they came across one another. The strengthening effect of the primary γ' precipitates could be reflected by the fact that the γ/γ' interfaces provided critical sites for hindering dislocation slip and SF propagation; it was found that scarce dislocations or SFs nucleated in the interior of the primary γ' precipitates could pass across the γ/γ' interfaces into the γ matrix.

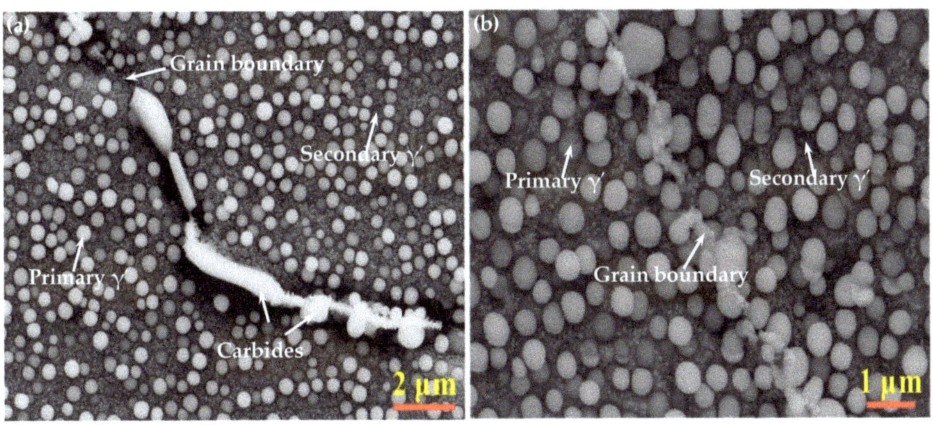

Figure 1. Typical microstructures of the samples observed by optical microscope: (**a**) original sample (ε = 0); (**b**) compressed sample (ε = 1.0).

The deformation microstructures of the tested sample at medium strains are depicted in Figure 3, which shows significant differences with respect to prior deformation at lower strains. When deformed at the strain of 0.3, numerous MT bundles were introduced that prevailed over the dislocation-controlled deformation and propagated parallelly with smaller spacing (Figure 3a). Meanwhile, SFs that extended along different directions were found to intersect with MTs, resulting in the formation of blocky obstacles, which would hinder dislocation motion and increase the density of dislocation tangle around these MTs. With the accumulation of dislocations due to the interaction between dislocations, SFs, and MTs, distorted stripes were introduced which had nearly straight boundaries that ran parallel to the MT boundaries, indicating the crucial role that MTs played on the formation of these distorted regions (Figure 3b). When deformed at the strain of 0.4, abundant distorted blocks emerged from the channels between the primary γ' precipitates, and the size of the distorted regions grew obviously (Figure 3c). Simultaneously, the density of dislocation in the γ matrix decreased remarkably due to long-term dislocation recovery, which was much more advantageous at higher strain. Under higher magnification, it could be found that these distorted blocks were sheared by fine SFs that propagated along two or more directions (Figure 3d). Furthermore, the interfaces of the primary γ' precipitates and γ matrix still had an impediment effect for SF shearing.

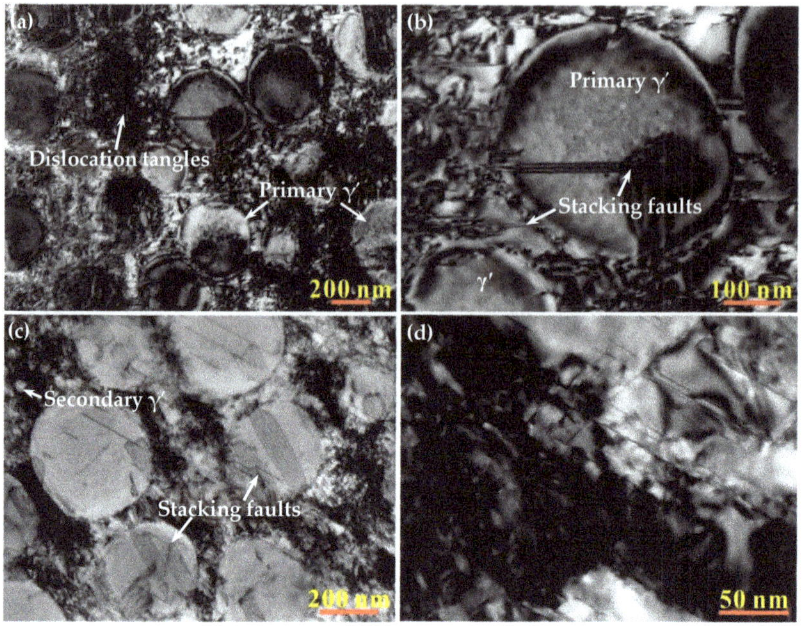

Figure 2. Typical microstructures of the compressed specimens at lower strains: (**a**) dislocation tangles, ε = 0.1; (**b**) SF nucleation, ε = 0.1; (**c**) SF propagation, ε = 0.2; (**d**) interaction between dislocations and SFs, ε = 0.2.

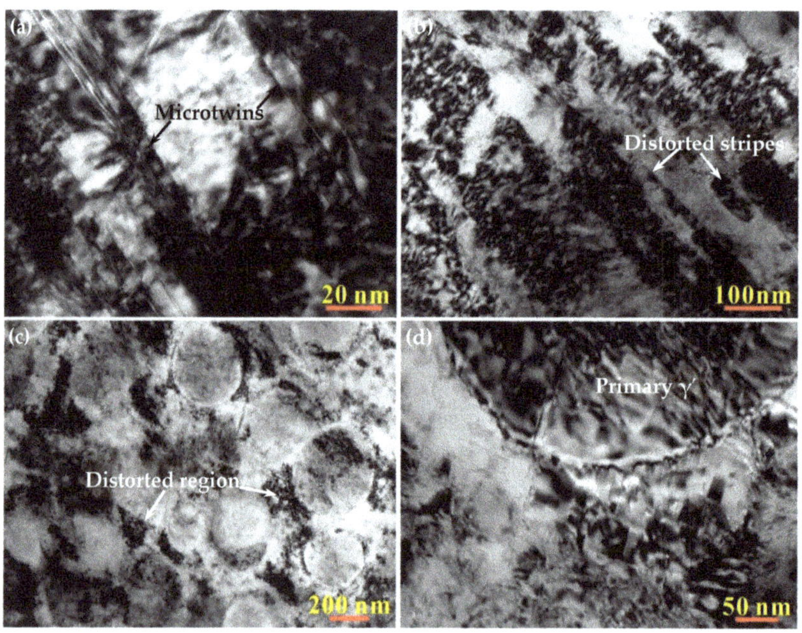

Figure 3. Typical microstructures of the compressed specimens at medium strains: (**a**) MT formation, ε = 0.3; (**b**) strip-like distorted region, ε = 0.3; (**c**) distorted blocks, ε = 0.4; (**d**) the γ/γ' interface, ε = 0.4.

The deformation microstructures of the tested sample at higher strains were depicted in Figure 4. When deformed at the strain of 0.6, the primary γ' precipitates were surrounded by numerous intersected SFs/MTs; these SFs/MTs seemed to penetrate the phase boundaries of primary γ' precipitates on account of the continuous dislocation activities (Figure 4a). Apart from the numerous distorted regions, some SGs were found in the localized zone (Figure 4b). These SGs could be classified into two types, which included polygonal SGs with straight GBs and banded SGs that arranged parallelly. When the compressive test was carried out at the strain of 1.0, abundant SGs with polygonal shape formed that replaced the original distorted dark regions (Figure 4c). Besides these SGs, numerous nanograins (NGs) could be detected in the forms of equiaxial or irregular particles, which was verified by the presence of diffraction rings using selected area electron diffraction (Figure 4d). It was shown that SFs/MTs and SGs/NGs could hardly coexist since SFs/MTs were constantly absent from the regions that had plentiful SGs or NGs when deformation occurred at true strain of 0.6 and 1.0.

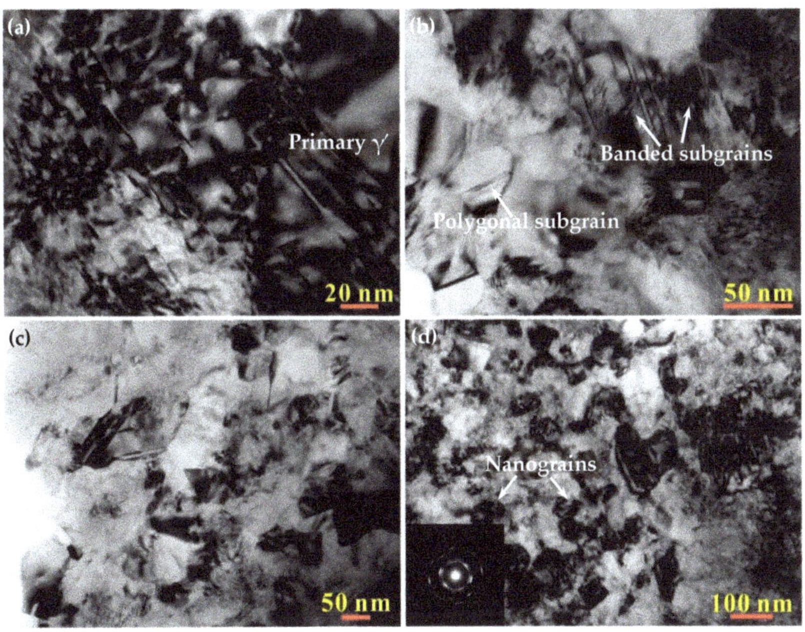

Figure 4. Typical microstructures of the compressed specimens at higher strains: (**a**) shearing of primary γ' precipitates by SFs/MTs, ε = 0.6; (**b**) SG formation, ε = 0.6; (**c**) SG refinement, ε = 1.0; (**d**) NG formation, ε = 1.0.

4. Discussion

It has been well documented that dislocation slipping dominates the high-temperature deformation for superalloys with higher SFE, while SF shearing or microtwinning can be motivated in superalloys with lower SFE [19–21]. The tested superalloy has relatively lower SFE by adding 20% Co element in order to obtain a high concentration of electron hole. By interrupted compressive deformation investigation, it is interesting to find that the deformation micromechanisms of the tested superalloy underwent continual transformation with the implied strain level, which was dominated by dislocation slipping at lower strain, then SF/MT shearing at medium strain, and SG/NG forming at higher strain. It is worth noting that oxygen atoms have obvious impacts on the deformation behaviors of the tested samples since the deviation from stoichiometry and appearance of the oxygen anions can lead to some changes in the charge state of the cations, which in turn will greatly change the electronic parameters [22,23]. That will seriously affect the practical application of the tested materials. It is well known that the complex transition metals and alloys easily allow the oxygen excess and/or deficit. In order to eliminate the adverse effects of oxygen atoms, the compression tests were carried out in a vacuum.

In this study, when the samples were deformed at lower strain and 725 °C, {111} <110> slip system was activated subsequently, then the a/2 <110> full dislocations dissociated into a/3 <112> Frank partial dislocation and a/6 <112> Shockley partial dislocation to facilitate the dislocation motion in face-centered cubic (FCC) metals [24]. Though the slipping dislocations could pass across fine secondary γ' precipitates by a shearing mechanism, they would be impeded and piled up by larger primary γ' precipitates unless dislocation climbing was activated. With the continuous slipping and climbing, most of the dislocations tangled around the γ/γ' interfaces, which transformed into obstacles for the movement of other dislocations. Subsequently, slip of paired a/6 <112> Shockley partial dislocations gradually turned to prevail over the individual dislocation motion, which led to the initiation of SF debris in the γ matrix and primary γ' precipitates. Further deformation led to the multidirectional initiation and propagation of SFs and accelerated the formation of dense dislocation tangles.

When compressive deformation occurred at medium strains, microtwinning was activated that produced numerous MT bundles. MT is a kind of special deformation twin that has a thickness of 4–50 atom layers, which play an important role in the deformation mechanisms of low SFE superalloys. It is reported that MTs could synchronously improve the strength and plasticity by acting both as dislocation blockers and dislocation slip planes in Ni-Co-based superalloys during tensile tests [13]. It has been well documented that MTs were generally introduced by severe plastic deformation that applied high strain rate and deformation amount at a low temperature [25,26]. However, this study shows that MTs can be introduced in Ni-Co-based superalloy during compressive deformation at a lower strain rate of 0.01 s^{-1} and a higher temperature of 725 °C, which is in accordance with other superalloys [27,28].

The underlying mechanisms for MT formation in the precipitation strengthening superalloys can be rationalized by a diffusion-controlled atom reordering theory [29,30]. This indicates that pseudo twins act as the critical prerequisite for MT formation, and that these are produced by the pairwise passage of four identical a/6 <112> Shockley partial dislocations along adjacent {111} slip planes. However, the pseudo twin has high-energy Al/Al nearest neighbor bonds in its complex stacking structure that are unstable in thermodynamics due to the relatively high anti-phase boundary (APB) energy in the LI_2-structured γ' precipitate. When deformation is carried out at a high temperature (650–800 °C) and low strain rate, the transformation of pseudo twins into true MTs will be thermally activated by atom reordering that eliminates the high-energy Al/Al atom bonds after irreversible atom diffusion and exchange [30]. Once MTs are introduced, MT boundaries will act as the barriers for dislocation motion and result in dislocation tangle. Meanwhile, the intersection of MTs and SFs tends to aggravate the impediment effect for dislocation motion by locking dislocation slip. With the great increase in dislocation density at MT boundaries, distorted stripes are evolved from the highly tangled dislocations and detwinned MTs. Then, the distorted stripes will develop into larger distorted regions with irregular shape during further deformation.

Under higher strain level, polygonal and banded SGs appear initially, and then transform into equiaxial or irregular NGs as the further deformation is applied. When compressive deformation is carried out at the strain of 0.6, the distorted regions are enlarged to accommodate more dislocations. With the rotation of deformed grains, the orientation difference between the distorted regions and surrounding matrix increases gradually, which changes into SGs as a result of severe deformation. Then, microtwinning will be activated in the newly formed SGs when the orientation difference is high enough to hinder dislocation motion and MT shearing into the γ matrix according to previous results [14]. Subsequently, the SGs are sheared and divided into finer SGs by MT cutting, which results in the formation of polygonal and banded SGs. With the continual refinement effects of MTs in these SGs, the grain size of SGs decreases extremely and NGs will be introduced finally. It is shown that MT formation serves as the precursor for nanocrystallization in the tested superalloys since NGs are absent from the procedure without MTs and are introduced after the formation of MTs. It can be rationalized that initial deformation is necessary to supply numerous distorted regions for SGs formation, and microtwinning generally prevails over dislocation slipping when dislocation motion is impeded in the distorted regions. MTs play a critical role in subdividing and fragmenting the SGs, which serve as the precursor for NG generation.

5. Conclusions

In order to understand the nanocrystallization mechanisms of the newly developed Ni-Co-based superalloys, compressive tests were carried out at 725 °C and 0.01 s^{-1} with different true strains. Three main conclusions can be drawn as follows:

(1) The deformation mechanisms of the tested superalloy evolve gradually with the applied compression strain, which in our study was controlled by dislocation slipping and SF shearing at lower strains, transformed into deformation microtwinning at medium strains, and then dominated by SG and NG formation at higher strains.
(2) A nanocrystallization approach is found in an Ni-Co-based superalloy with low SFE via compression deformation at a high temperature and low strain rate, which exhibits obviously differently from the high-SFE alloys that NG formation creates at a low temperature and high strain rate, revealing that the NG formation of the tested superalloy is a thermal activation-assisted process of microstructural evolution.
(3) The NG formation of the tested superalloy can be ascribed to high strain and microtwinning. High strain applied during compression tests produces plentiful dislocations and distorted micro-sized SGs; then, microtwinning resulting from low SFE will refine the micro-sized SGs into NGs with further deformation.

Author Contributions: Investigation—H.X., L.W., X.L.; writing—Original draft preparation, H.X.; writing—Review and editing, Z.L.; supervision Z.Z., J.W., Y.G. All authors have read and agreed to the published version of the manuscript.

Funding: This work was supported by the Key Project Program (No. CHDKJ200110) of China Huadian Corporation LTD, Beijing.

Institutional Review Board Statement: Not applicable.

Informed Consent Statement: Not applicable.

Data Availability Statement: The data of this study are available from the corresponding author upon reasonable request.

Acknowledgments: The authors are also grateful to the Material Service Behavior Division of Institute of Metal Research.

Conflicts of Interest: The authors declare no conflict of interest.

References

1. Huang, K.J.; Li, W.; Pan, K.; Lin, X.; Wang, A.H. Microstructure and corrosion properties of $La_2Zr_2O_7$/NiCoAlY thermal barrier coatings deposited on Inconel 718 superalloy by laser cladding. *Coatings* **2021**, *11*, 101–112. [CrossRef]
2. Antonov, S.; Rozman, K.A.; Jablonski, P.D.; Detrois, M. Implications of carbon content for the processing, stability, and mechanical properties of cast and wrought Ni-based superalloy Nimonic 105. *Mater. Sci. Eng. A* **2022**, *857*, 144049. [CrossRef]
3. Kozlovskiy, A.L.; Zdorovets, M.V. Synthesis, structural, strength and corrosion properties of thin films of the type CuX (X = Bi, Mg, Ni). *J. Mater. Sci. Mater. Electron.* **2019**, *30*, 11819–11832. [CrossRef]
4. Shlimas, D.I.; Kozlovskiy, A.L.; Zdorovets, M.V. Study of the formation effect of the cubic phase of $LiTiO_2$ on the structural, optical, and mechanical properties of $Li_{2\pm x}Ti_{1\pm x}O_3$ ceramics with different contents of the X component. *J. Mater. Sci. Mater. Electron.* **2021**, *32*, 7410–7422. [CrossRef]
5. Gu, Y.F.; Cui, C.Y.; Yuan, Y.; Zhong, Z.H. Research progress in a high performance cast & wrought superalloy for turbine disc applications. *Acta Metall. Sin.* **2015**, *51*, 1191–1206.

6. Gu, Y.F.; Fukuda, T.; Cui, C.Y.; Harada, H.; Mitsuhashi, A.; Yokokawa, T.; Fujioka, J.; Koizumi, Y.; Kobayashi, T. Comparison of mechanical properties of TMW alloys, new generation of cast-and-wrought superalloys for disk applications. *Metall. Mater. Trans. A* **2009**, *40*, 3047–3050. [CrossRef]
7. Gu, Y.F.; Harada, H.; Cui, C.Y.; Ping, D.H.; Sato, A.; Fujioka, J. New Ni-Co-base disk superalloys with higher strength and creep resistance. *Scripta Mater.* **2006**, *55*, 815–818. [CrossRef]
8. Cui, C.Y.; Gu, Y.F.; Ping, D.H.; Harada, H. Microstructural Evolution and Mechanical Properties of a Ni-Based Superalloy, TMW-4. *Metall. Mater. Trans. A* **2009**, *40*, 282–291. [CrossRef]
9. Luo, Z.P.; Wu, Z.T.; Miller, D.J. The dislocation microstructure of a nickel-base single-crystal superalloy after tensile fracture. *Mater. Sci. Eng. A* **2003**, *354*, 358–368. [CrossRef]
10. Tian, S.G.; Guo, Z.G.; Yu, H.C.; Xue, Y.C.; Zhang, S.; Su, Y.; Meng, F.L. Microstructure evolution and FEM analysis of a [0 1 1] oriented single crystal nickel-based superalloy during compressive creep. *J. Alloys Compd.* **2013**, *563*, 135–142. [CrossRef]
11. Yuan, Y.; Gu, Y.F.; Cui, C.Y.; Osada, T.; Zhong, Z.H.; Tetsui, T.; Yokokawa, T.; Harada, H. Influence of Co content on stacking fault energy in Ni–Co base disk superalloys. *J. Mater. Res.* **2011**, *216*, 2833–2837. [CrossRef]
12. Tian, C.G.; Han, G.M.; Cui, C.Y.; Sun, X.F. Effects of Co content on tensile properties and deformation behaviors of Ni-based disk superalloys at different temperatures. *Mater. Des.* **2015**, *88*, 123–131. [CrossRef]
13. Xu, H.; Zhang, Z.J.; Zhang, P.; Cui, C.Y.; Jin, T.; Zhang, Z.F. The synchronous improvement of strength and plasticity (SISP) in new Ni-Co based disc superalloys by controlling stacking fault energy. *Sci. Rep.* **2017**, *7*, 8046. [CrossRef] [PubMed]
14. Xu, H.; Zhang, Z.J.; Zhang, P.; Cui, C.Y.; Jin, T.; Zhang, Z.F. Formation of nanograins in Ni-Co based superalloys compressed quasistatically at high temperature. *Scr. Mater.* **2017**, *136*, 92–96. [CrossRef]
15. Li, W.; Liu, P.; Ma, F.C.; Rong, Y.H. Microstructural characterization of nanocrystalline nickel produced by surface mechanical attrition treatment. *J. Mater. Sci.* **2009**, *44*, 2925–2930. [CrossRef]
16. Tao, N.R.; Wang, Z.B.; Tong, W.P.; Sui, M.L.; Lu, J.; Lu, K. An investigation of surface nanocrystallization mechanism in Fe induced by surface mechanical attrition treatment. *Acta Mater.* **2002**, *50*, 4603–4616. [CrossRef]
17. Zhao, W.S.; Tao, N.R.; Guo, J.Y.; Lu, Q.H.; Lu, K. High density nano-scale twins in Cu induced by dynamic plastic deformation. *Scr. Mater.* **2005**, *53*, 745–749. [CrossRef]
18. Lu, K.; Lu, J. Surface nanocrystallization of metallic materials-presentation of the concept behind a new approach. *J. Mater. Sci. Technol.* **1999**, *15*, 193–197.
19. Yuan, Y.; Gu, Y.F.; Osada, T.; Zhong, Z.H.; Yokokawa, T.; Harada, H. A new method to strengthen turbine disc superalloys at service temperatures. *Scr. Mater.* **2012**, *66*, 884–889. [CrossRef]
20. Gopinath, K.; Gogia, A.K.; Kamat, S.V.; Balamuralikrishnan, R.; Ramamurty, U. Tensile Properties of Ni-Based Superalloy 720Li: Temperature and Strain Rate Effects. *Metall. Mater. Trans. A* **2008**, *39*, 2340–2350. [CrossRef]
21. Li, Y.; Hong, Z.J.; Liu, B.G.; Jia, X.S. Investigation of microstructure evolution and mechanical properties of multi-precipitation Ni-Co base superalloys. *Mater. Sci. Eng. A* **2021**, *801*, 140333. [CrossRef]
22. Trukhanov, S.V.; Trukhanov, A.V.; Vasiliev, A.N.; Balagurov, A.M.; Szymczak, H. Magnetic state of the structural separated anion-deficient $La_{0.70}Sr_{0.30}MnO_{2.85}$ manganite. *J. Exp. Theor. Phys.* **2011**, *113*, 819–825. [CrossRef]
23. Kozlovskiy, A.; Egizbek, K.; Zdorovets, M.V.; Ibragimova, M.; Shumskaya, A.; Rogachev, A.A.; Ignatovich, Z.V.; Kadyrzhanov, K. Evaluation of the efficiency of detection and capture of manganese in aqueous solutions of FeCeOx nanocomposites doped with Nb_2O_5. *Sensors* **2020**, *20*, 4851. [CrossRef] [PubMed]
24. Zhong, Z.H.; Gu, Y.F.; Yuan, Y.; Osada, T.; Cui, C.Y.; Yokokawa, T.; Tetsui, T. Effect of solution temperature on the microstructure and mechanical properties of a newly developed superalloy TMW-4M3. *Metall. Mater. Trans. A* **2012**, *43*, 1017–1025. [CrossRef]
25. Qu, S.; An, X.H.; Yang, H.J.; Huang, C.X.; Yang, G.; Zang, Q.S.; Wang, Z.G.; Wu, S.D.; Zhang, Z.F. Microstructural evolution and mechanical properties of Cu-Al alloys subjected to equal channel angular pressing. *Acta Mater.* **2009**, *57*, 1586–1601. [CrossRef]
26. Hong, C.S.; Tao, N.R.; Huang, X.; Lu, K. Nucleation and thickening of shear bands in nano-scale twin/matrix lamellae of a Cu-Al alloy processed by dynamic plastic deformation. *Acta Mater.* **2010**, *58*, 3103–3116. [CrossRef]
27. Viswanathan, G.B.; Sarosi, P.M.; Henry, M.F.; Whitis, D.D.; Milligan, W.W.; Mills, M.J. Investigation of creep deformation mechanisms at intermediate temperatures in René 88 DT. *Acta Mater.* **2005**, *53*, 3041–3057. [CrossRef]
28. Knowles, D.M.; Chen, Q.Z. Superlattice stacking fault formation and twinning during creep in γ/γ' single crystal superalloy CMSX-4. *Mater. Sci. Eng. A* **2003**, *340*, 88–102. [CrossRef]
29. Kolbe, M. The high temperature decrease of the critical resolved shear stress in nickel-base superalloys. *Mater. Sci. Eng. A* **2001**, *319–321*, 383–387. [CrossRef]
30. Kovarik, L.; Unocic, R.R.; Li, J.; Sarosi, P.; Shen, C.; Wang, Y.; Mills, M.J. Microtwinning and other shearing mechanisms at intermediate temperatures in Ni-based superalloys. *Prog. Mater. Sci.* **2009**, *54*, 839–873. [CrossRef]

Disclaimer/Publisher's Note: The statements, opinions and data contained in all publications are solely those of the individual author(s) and contributor(s) and not of MDPI and/or the editor(s). MDPI and/or the editor(s) disclaim responsibility for any injury to people or property resulting from any ideas, methods, instructions or products referred to in the content.

coatings

MDPI

Review

Application of Laser Welding in Electric Vehicle Battery Manufacturing: A Review

Junbo Feng [1,2], Peilei Zhang [1,2,*], Hua Yan [1,2], Haichuan Shi [1,2], Qinghua Lu [1,2], Zhenyu Liu [1,2], Di Wu [1,2], Tianzhu Sun [3], Ruifeng Li [4] and Qingzhao Wang [5]

1 School of Materials Science and Engineering, Shanghai University of Engineering Science, Shanghai 201620, China; 18990122132@163.com (J.F.); yanhua@foxmail.com (H.Y.); shc0010@126.com (H.S.); luqh@sues.edu.cn (Q.L.); zhenyu_ender@foxmail.com (Z.L.); wudi612@126.com (D.W.)
2 Shanghai Collaborative Innovation Center of Laser Advanced Manufacturing Technology, Shanghai 201620, China
3 Warwick Manufacturing Group (WMG), University of Warwick, Coventry CV4 7AL, UK; tianzhu.sun@warwick.ac.uk
4 School of Materials Science and Engineering, Jiangsu University of Science and Technology, Zhenjiang 212003, China; li_ruifeng@just.edu.cn
5 SANY Heavy Industry Co., Ltd., Shanghai 201306, China; wangrain9_14@163.com
* Correspondence: peilei@sues.edu.cn

Abstract: Electric vehicle battery systems are made up of a variety of different materials, each battery system contains hundreds of batteries. There are many parts that need to be connected in the battery system, and welding is often the most effective and reliable connection method. Laser welding has the advantages of non-contact, high energy density, accurate heat input control, and easy automation, which is considered to be the ideal choice for electric vehicle battery manufacturing. However, the metal materials used for the electrodes of the battery and the connectors used to connect the battery are not the same, so the different materials need to be welded together effectively. Welding different materials together is associated with various difficulties and challenges, as more intermetallic compounds are formed, some of which can affect the microstructure, electrical and thermal properties of the joint. Because the common material of the battery housing is steel and aluminum and other refractory metals, it will also face various problems. In this paper reviews, the challenges and the latest progress of laser welding between different materials of battery busbar and battery pole and between the same materials of battery housing are reviewed. The microstructure, metallographic defects and mechanical properties of the joint are discussed.

Keywords: electric vehicle battery; laser welding; welding defects; dissimilar metal; identical metal

1. Introduction

Due to global warming, today's climate problems are intensifying, and extreme weather is occurring frequently. The main cause of this problem group is greenhouse gas emissions, mainly carbon dioxide (90%), and the transport sector is one of the largest contributors to greenhouse gas emissions, according to the International Energy Agency (IEA), and in 2015, global CO_2 emissions reached 323 billion tons, while transport accounted for 24% of the total emissions. Three quarters of this is contributed by the road component [1]. To mitigate climate change, carbon emission laws have been enacted around the world [2,3]. New energy vehicles (NEV), as an alternative to traditional internal combustion engine vehicles (ICEV), are rapidly developing in major international automotive markets. China, the United States, Japan, Germany and other countries have restricted the sales of traditional internal combustion engine vehicles at the national level and formulated a series of new policies to encourage the development of new energy vehicles, so as to reduce the use of oil and reduce carbon dioxide emissions [4].

The fastest developing new energy vehicles are electric vehicles (EVs), which are powered by power batteries. Lithium-ion battery has become the most important power supply for electric vehicles because of its high energy density, low self-discharge and long life cycle [5–7]. Batteries used in electric vehicles are mainly small solid cylindrical batteries, large solid prismatic batteries and large soft bag or polymer batteries [8–10], as shown in Figure 1. Battery packs for electric vehicles are usually designed and manufactured in a battery-module-cell structure, as shown in Figure 2. The main difference in practice is how to achieve the required battery capacity and power. A small number of large capacity cells can be connected in series, as shown in Figure 2a. Alternatively, multiple small batteries with small capacity are connected in parallel and then connected in series to form high-capacity modules, as shown in Figure 2b. These batteries are usually connected by busbars [11], as shown in Figure 3. Power batteries usually work in harsh driving environments, such as vibration, high temperature and possible collision. How to securely connect hundreds of connections in battery modules is related to the new performance and safety of the entire battery system. Various bonding techniques, such as laser welding, friction stir welding, tungsten inert gas welding, ultrasonic lead bonding and resistance spot welding, have been used in battery manufacturing [8,10,12]. Ultrasonic welding mainly uses high-frequency vibration, usually 20 kHz or above, to connect materials by forming solid-state bonds under clamping pressure [13]. Ultrasonic welding is suitable for the welding of multiple thin foils, dissimilar materials or highly conductive materials. It is mainly used in banded batteries [14], and electric vehicle batteries are usually cylindrical or prismatic batteries, which may destroy the integrity of the battery structure under the action of pressure and vibration, so it is not suitable for the welding of electric vehicle batteries [8]. The working principle of resistance spot welding is mainly to apply pressure on the contact surface of the workpiece and connect large current to cause partial melting of the workpiece [12]. However, the commonly used materials in electric vehicle batteries are aluminum and copper, and aluminum and copper have the characteristics of high electrical and thermal conductivity, so resistance welding is difficult to weld. Laser welding is considered to be the most promising connection method because of its easy automation, high accuracy, small heat-affected zone, non-contact process, high process speed and ease of welding different metals. Laser welding is an efficient and precise welding method using high energy density laser beam as heat source. Due to heat concentration, fast welding speed, small thermal effect, small welding deformation, easy to realize efficient automation and integration [15–17], it is more and more widely used in power battery manufacturing.

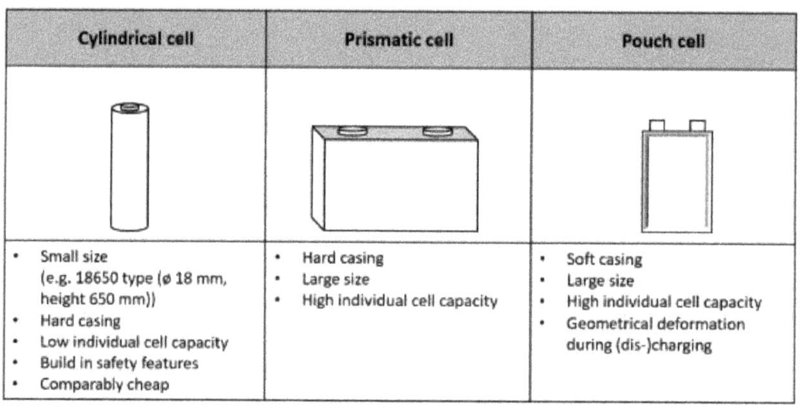

Figure 1. Overview of different cell types used in automotive battery applications: (**left**) cylindrical cell, (**middle**) prismatic cell, and (**right**) pouch cell. Reprinted with permission from Ref. [10]. Copyright 2020, Elsevier.

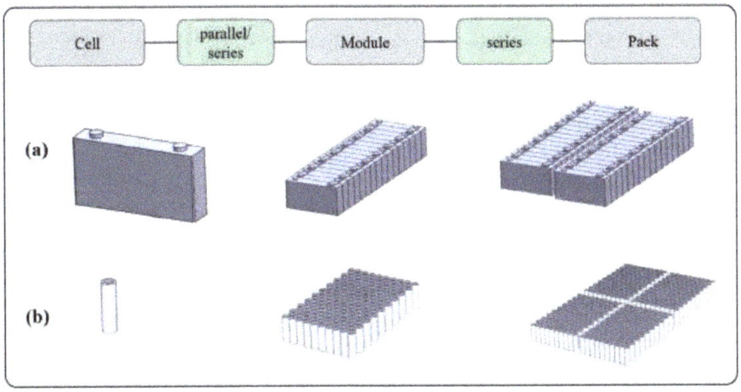

Figure 2. Overview of battery packs indicating two constructions with (**a**) cylindrical and (**b**) prismatic cells. Reprinted with permission from Ref. [10]. Copyright 2020, Elsevier.

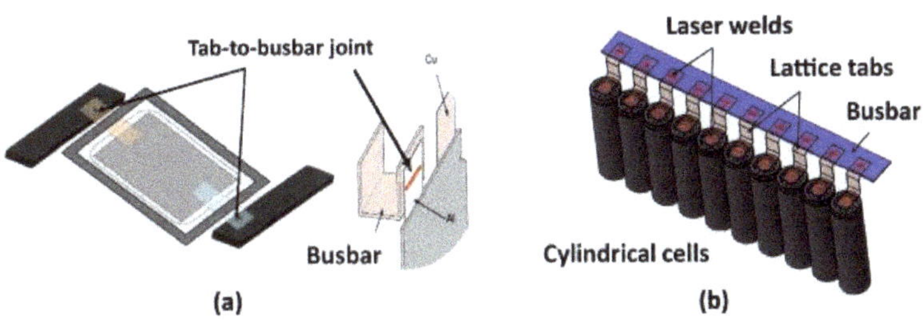

Figure 3. An illustration of tab-to-busbar joints made during (**a**) pouch cell-based and (**b**) cylindrical cell-based module manufacture [11].

In addition, the battery connection can be mechanically fastened in a variety of ways, including nut bolts, spring fasteners, screws or fasteners [18]. Nut and bolt joints may be either physically distinct nut and bolt assemblies or a threaded feature, for example, electrode and nut. For battery module level connections, nut and bolt joints are mainly limited to prismatic cells. Some special battery modules are not suitable for permanent connection (such as welding) due to the need for battery maintenance, so mechanical nuts and bolts can also be used for connection of special battery systems. At present, battery casings are mainly produced by welding sheets, so some welding defects, such as pores and cracks, are inevitable. Niu et al. [19] added high entropy alloy to aluminum powder during additive manufacturing of aluminum, which inhibited the crack generation and improved the strength. Therefore, with the development of laser additive, the battery case may be produced by additive manufacturing in the future.

This paper mainly reviews the laser welding of dissimilar metal joints between battery and bus in electric vehicle battery system, as well as the packaging of the same metal between battery pack by laser welding. The difficulties and challenges of laser welding between homogenous and dissimilar metals are discussed. Especially, the welding defects caused by the welding between dissimilar metals, the causes of these welding defects and the possible solutions are put forward.

2. Welding between Batteries and Busbars

In all the production processes of power battery packs, there is a key process, that is, the welding of a single lithium battery and the connector. This is the key to the quality of series and parallel lithium-ion battery cells, that is, the welding of the battery pole and the busbars. The quality of the welding here will directly affect the reliability of the quality of the lithium-ion battery pack used as a power source for electric vehicles. In addition, due to the relative particularity of lithium-ion battery, the welding technology has also put forward high requirements. If the welding strength is weak, the internal resistance of the battery string will increase, thus affecting the normal power supply of the battery string. Excessive welding heat will cause the electrode cover of the battery core to be penetrated, resulting in electrolyte leakage and battery circuit's short circuit, resulting in battery combustion or even explosion, which seriously threatens the safety of passengers and drivers. It is because of the problems of unreliable welding quality and low welding efficiency in series and parallel welding of power battery pack that the safety and production efficiency of power battery pack are very low. Therefore, in order to ensure the safety of its use and production efficiency, the welding between the battery pole and the busbars must be reliable, which is an important factor to ensure the product yield and service life.

2.1. Aluminum and Steel

Battery busbars are made of two common materials: copper and aluminum. The battery electrode materials are usually steel and aluminum, and the parameters and challenges of laser welding are different. Aluminum has the advantages of good electrical conductivity, light weight and good plasticity, so it is very suitable as a busbar material. The efficient and reliable connection of steel and aluminum can provide huge economic benefits for battery manufacturing.

However, the connection between the aluminum busbars and the steel poles of the battery is challenging because iron and aluminum have great differences in thermal physical properties such as melting point, thermal conductivity and thermal expansion coefficient. In addition, the low solubility between Fe and Al leads to the formation of brittle intermetallic layers where iron and aluminum are metallurgically incompatible, and the resulting fusion welding is prone to the formation of harmful intermetallic compounds (IMCs). The formation of intermetallic compounds (IMCs) has been shown to cause a variety of welding defects, such as microcracks and pores [20–24]. The chemical composition, crystal structure, hardness and Gibbs free energy of various IMCs formed in Fe–Al binary system are shown in Table 1 [25–27]. Fe_2Al_5, Fe_4Al_{13} and$FeAl_2$ are Al-rich phases, and FeAl and Fe_3Al are Fe-rich phases. As can be seen from Table 1, aluminum-rich IMCs have stronger hardness and brittleness than iron-rich IMCs, so cracks and other defects are more likely to occur between welded joints. The iron-rich IMCs have better toughness and ductility, which can reduce the generation of cracks. However, in terms of Gibbs free energy, the formation of the Al-rich phase is thermodynamically more favorable to the formation of the Fe-rich phase. Fe_2Al_5 is thermodynamically more stable, forming first, followed by Fe_4Al_{13}, $FeAl_2$, FeAl, and $FeAl_3$ [28]. IMCs are usually resistive, and too much IMCs will increase the internal resistance of the battery system, resulting in more Joule heat generated during the charging and discharging process of the battery system, affecting the life of the battery system. Therefore, the generation of IMC phase in the weld tissue should be controlled as much as possible during the welding process.

Table 1. Fe–Al IMC properties.

Phases	Al at%	Hardness, HV	Crystal Structure	ΔG (KJ mol^{-1})
Fe_2Al_5	70–73	1000–1100	orthorhombic	−19.64
$Fe_4Al_{13}/FeAl_3$	74.5–76.6	820–980	BC monoclinic	−22.87
FeAl2	66–66.9	1000–1050	triclinic	−17.0
FeAl	23–55	400–520	Simple cubic (B2 type)	−11.09
Fe_3Al	23–34	250–350	FCC	−4.83

Yang et al. [29] found that when welding aluminum and steel, the penetration depth should be controlled within a certain range, and Fe–Al IMCs rich in iron are mainly formed when the penetration depth was low. When the penetration depth was increased, the aluminum rich Fe–Al IMCs were mainly formed, which would make the mechanical properties of the joint worse. Chen et al. [30] added magnetic field action perpendicular to the welding direction when welding 301 stainless steel and 5754 aluminum alloy. They found that increasing the magnetic field inhibited the diffusion of C atoms and reduced the austenite grain size. Increasing the magnetic field could also effectively inhibit the concentration of Al in the joint, as shown in Figure 4, thus reducing the cracks and thickness of IMCs at the interface, improving the shear strength of the joint and reducing the hardness of the joint.

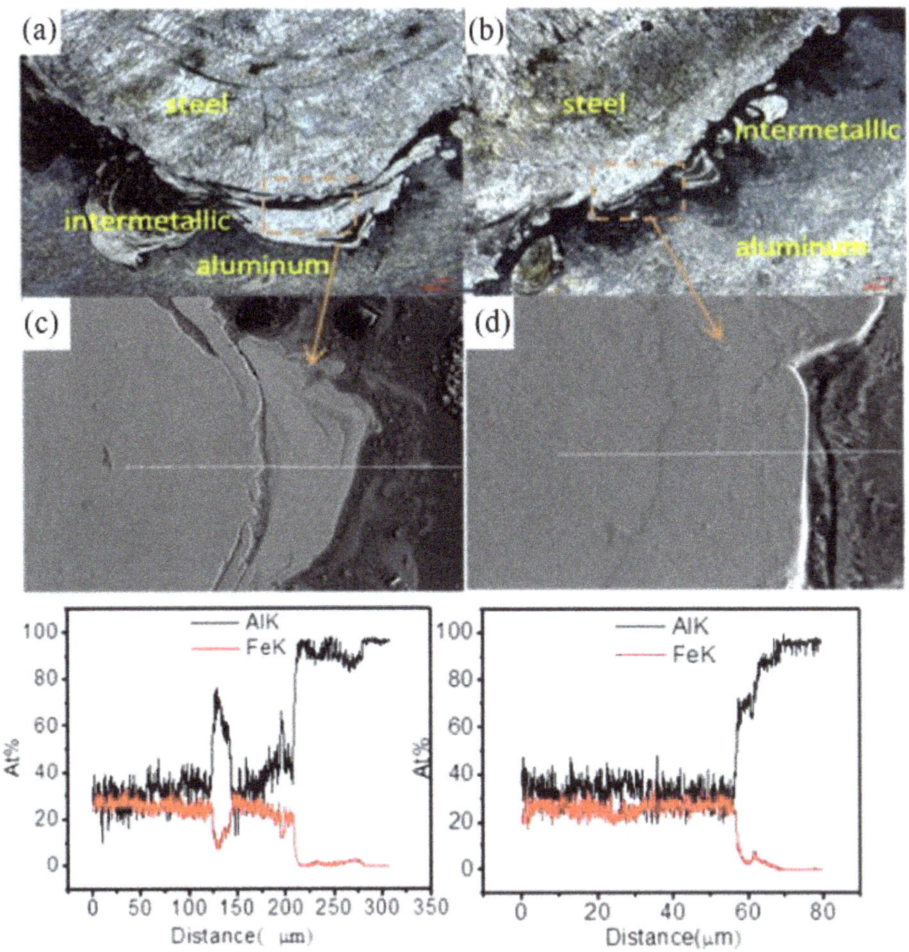

Figure 4. Morphology of interface of two groups: (**a**) B = 0 and (**b**) B = 240 mT; Line scanning area and analysis of the seam–aluminum interface layer: (**c**) B = 0 and (**d**) B = 240 mT. Reprinted with permission from Ref. [30]. Copyright 2016, Elsevier.

Torkamany et al. [31] welded 0.8 mm thick mild steel (st14) with a 2 mm thick 5754 aluminum alloy. They found that when the power of the pulsed laser was too high, it was not conducive to the formation of the weld. When the laser power was higher, it increased the mixing of steel and aluminum, increased the content of aluminum in the weld and formed more intermetallic compounds. Increasing the duration of the laser pulse would have a similar effect with the increase in the amount of heat input. These regions with more intermetallic compounds form cracks under thermal stress, as shown in Figure 5. On the other hand, reducing the duration of the laser pulse below a critical level led to a lack of fusion. Increasing the welding speed would also lead to incomplete interface fusion and reduce joint strength. They reported optimal values of process parameters for producing high-strength welds due to low intermetallic compound content, high surface quality and no obvious defects in the continuous interface layer. The peak power was 1430 W, the pulse duration is 5 ms, and the welding speed was 4 mm/s.

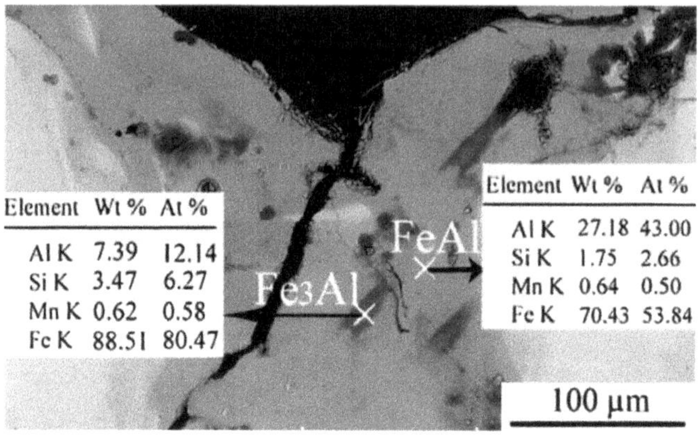

Figure 5. SEM micrograph around the joint crack. Reprinted with permission from Ref. [31]. Copyright 2010, Elsevier.

Chen et al. [32] studied the effect of intermediate nickel foil layer on welding of A5052 aluminum alloy and 201 stainless steel. The images with and without nickel layers are shown in Figure 6a,b. In the weld with nickel layer, an intermetallic layer could be obviously seen, which can be divided into $FeAl_3$ and $Al_{0.9}Ni_{1.1}$ layers, as shown in Figure 6c, indicating that nickel foil changed the composition of intermetallic compounds. They believed that because Al had a certain solubility in α-Fe, when aluminum was mixed into the molten steel as a solute element, no intermetallic compounds were observed in the fusion zone. The welding depth had a significant effect on the mechanical properties of the weld. Initially, the tensile strength increased as the welding depth reaches 300 μm, but as the welding depth further increased, the tensile strength began to decrease, which was due to the higher aluminum content, forming a more brittle intermetallic compound. In addition, tensile tests and microhardness measurements of welded samples showed that Ni foil increased the tensile strength while reducing the microhardness of the intermetallic layer.

Cao et al. [33] shifted the focus of laser welding to the stainless steel side when welding aluminum alloy and stainless steel. They found that a certain amount of laser offset could effectively improve the tensile strength of the weld. When the laser offset was 0.2 mm, the tensile strength of the welded joint reaches 129.6 MPa. Wei et al. [34] found that when welding SUS3010S stainless steel and 6061 aluminum alloy, laser cleaning of the stainless steel layer could effectively enhance the bonding strength of the welded joint. They used the laser to clean the stainless steel at the same time, so that the stainless steel surface under the action of the laser to form a fish scale pit. They found that the molten Al could spread and fill the scale pits on these stainless steel surfaces, resulting in a significant increase in the mechanical strength and tensile shear resistance of the lap joints, as shown in Figure 7. After surface cleaning of stainless steel at 5000 mm/s and 1064 W laser power with optimal parameters, the tensile shear force of the weld after welding was increased by 54%.

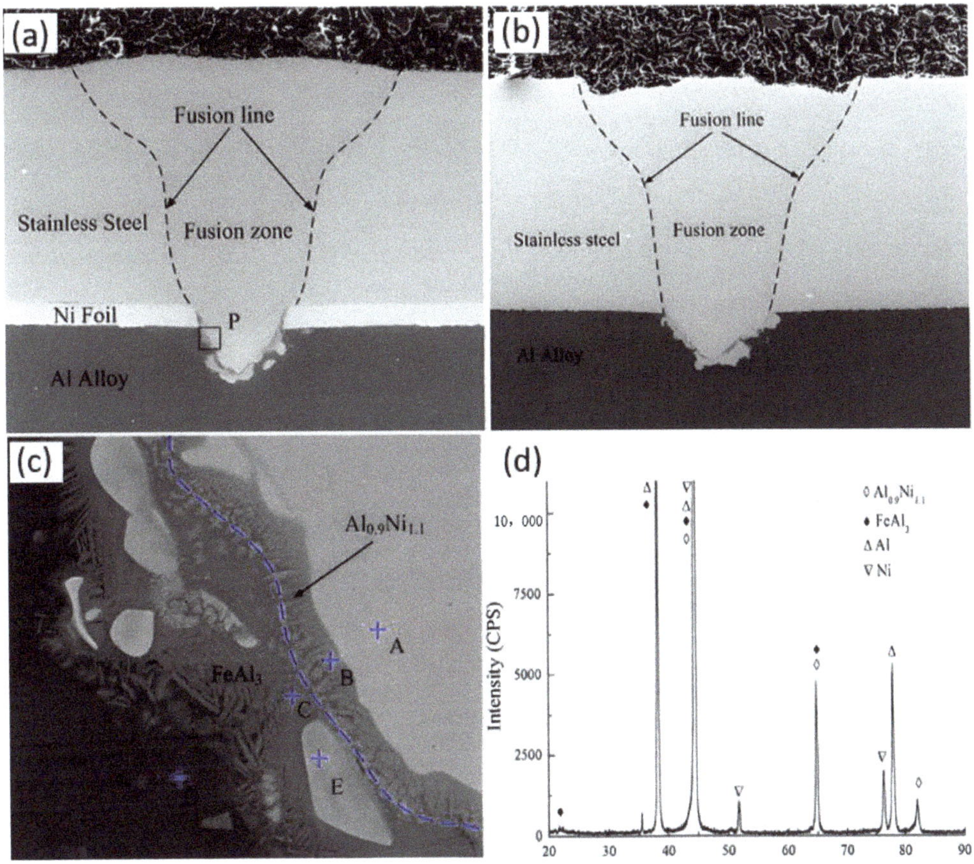

Figure 6. Microstructures of stainless steel/aluminum alloy joint. (**a**) Cross-section of the joint with Ni foil, (**b**) cross-section of the joint without Ni foil, (**c**) interfacial microstructures and (**d**) XRD patterns at the reaction zones. Reprinted with permission from Ref. [32]. Copyright 2012, Elsevier.

Chelladurai et al. [35] studied the welding of 3 mm 1050 aluminum and 0.25 mm steel sheet. The weld adopted the overlapping structure, the top was thin nickel steel, and the bottom was thick aluminum. They found that in laser welding, adding a wobble to the beam could effectively improve the shape of the welded joint. When the wobble amplitude was 0.2, 0.4 and 0.6 mm, there were visible cracks on the surface; when the wobble amplitude was 0.8–1.2 mm, there were slight spatters on the weld but almost no cracks on the weld surface, as shown in Figure 8; when the wobble amplitude was 0.2–0.4 mm, serious cracks appear, while the wobble amplitude of 0.6 and 0.8 mm shows slight cracks. When the wobble amplitude was 1 and 1.2 mm, there was no crack in the weld, as shown in Figure 9. A small wobble weld (<0.4 mm) showed a high degree of mixing of iron and aluminum, resulting in a large IMC phase forming in the top and middle regions. The FeAl phase mainly existed in the top, steel side and some rich Al Fe_4Al_{13} on the weld/Al side. When the wobble amplitude was greater than 0.6 mm, the top of the weld is mainly Fe solid solution structure, and there was a small amount of Al, Fe_4A_{13} and Fe_2Al_5 phase IMC at the bottom of the weld. When the wobble amplitude was greater than 0.8 mm, the melting area was larger and the formed weld had fewer IMC phases, so there was lower resistance. On the contrary, small wobble welds exhibited higher resistance because of more IMC phases, less contact area and more cracks.

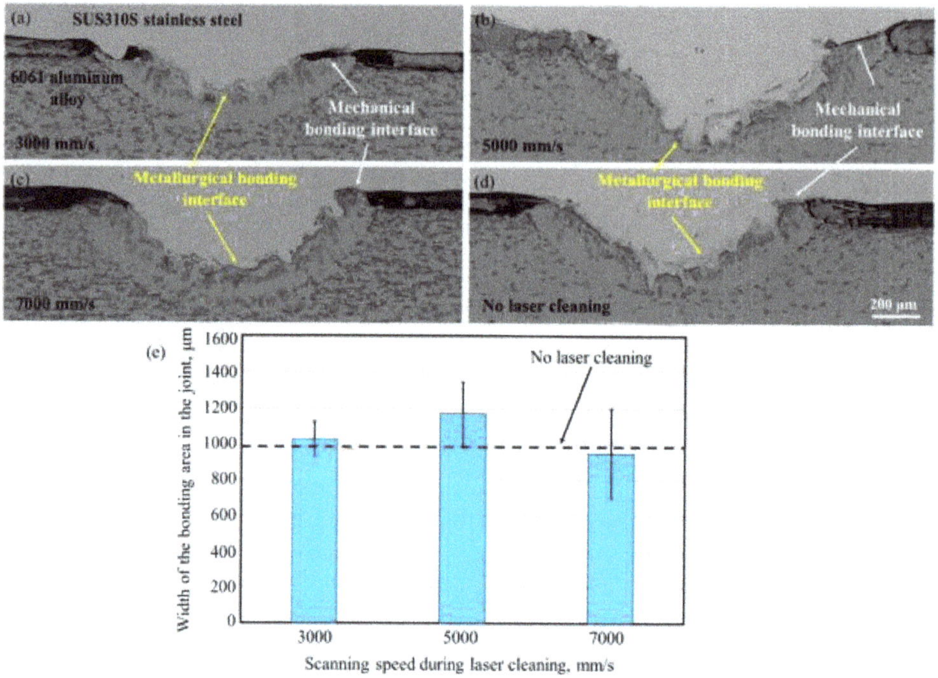

Figure 7. (**a–d**) Interface morphology at scanning speeds of 3000 mm/s, 5000 mm/s, 7000 mm/s and 0, respectively during laser cleaning. (**e**) Length of bonding area in the welded joint at different scanning speeds. Reprinted with permission from Ref. [34]. Copyright 2021, Elsevier.

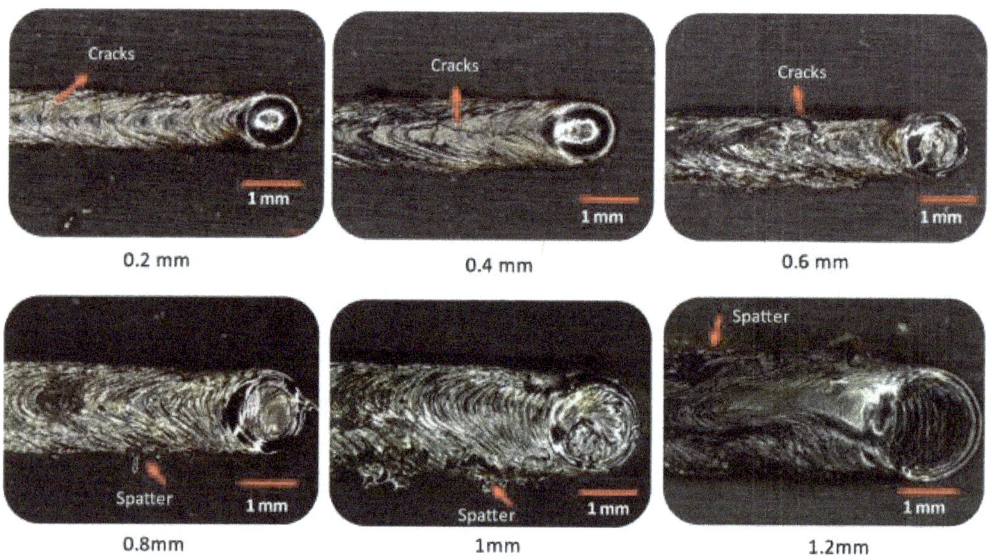

Figure 8. Surface images of welds of wobble amplitudes 0.2 to 1.2 mm [35].

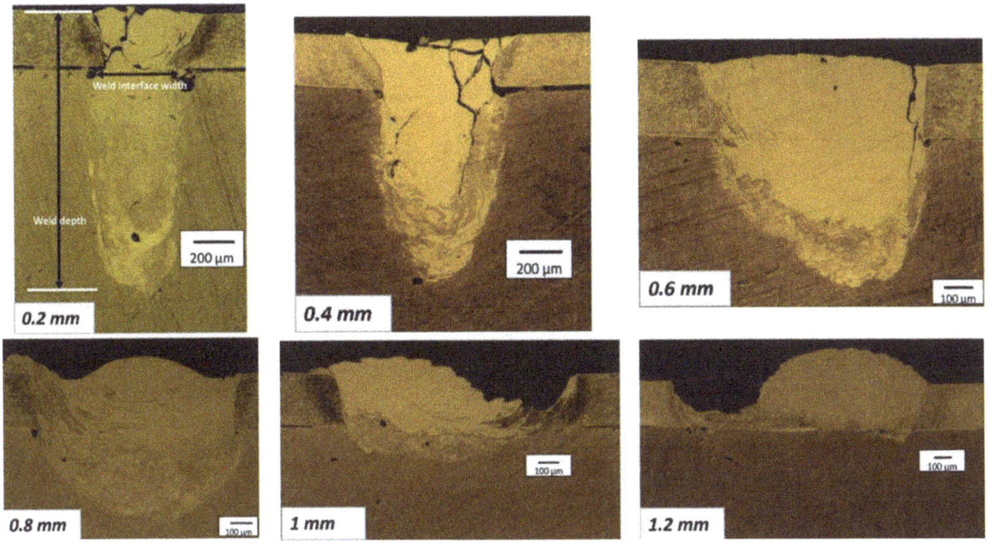

Figure 9. Weld macrostructures in different wobble amplitudes of 0.2 mm to 1.2 mm [35].

In order to improve the mechanical properties of aluminum and steel joints during welding, Sierra et al. [36] added Al-12Si for laser wire filling welding when welding AA6016 and DC04 steel, and they found that Si had a certain effect on the growth of Fe–Al intermetallic compounds. Fe–Al–Si intermetallic compounds with better mechanical properties were formed in the weld. Zhang et al. [37] also found that when laser welding H220YD steel and AA6016 aluminum used Al-5Si containing Si as interlayer, Al_8Fe_2Si, θ-$Al_{13}Fe_4$ and ξ-Al_2Fe intermetallic compounds of a certain thickness could be formed. When the thickness is greater than 10 µm, the joint strength decreases. Xia et al. [38] found that when welding 6061-T6 aluminum and DP590 steel, adding a sandwich containing Si could effectively reduce the laser power required for laser welding, and the formed intermetallic compounds containing Si had higher strength and shape, which could effectively improve the weld performance.

Reviewing the research in recent years, the laser welding of aluminum and steel has made great technical progress, but there is still a distance from the actual large-scale wide application, mainly because the mechanical properties of the joint are still insufficient. Progress has been made in process parameters and welding methods, but the formation of intermetallic compounds still needs to be solved, and there are other welding defects that need to be solved, such as pores and cracks. Table 2 summarizes the research on laser welding of aluminum and steel dissimilar metals.

2.2. Copper and Aluminum

Due to the differences in the melting point, thermal conductivity and thermal expansion of the two metals, the welding of copper–aluminum joints poses a major challenge [39–43]. Table 3 shows the main intermetallic compounds that can be formed between Cu and Al. Cu and Al in the welding process can form Cu_2Al, Cu_4Al_3, CuAl, Cu_9Al_4 and other intermetallic compounds [44]. The formation of these intermetallic compounds will greatly affect the microstructure and mechanical properties of the weld between Cu and Al [40,45–48]. Heideman et Al. [45] found that Cu and AL were welded with friction stir welding. With friction stir welding, which had the characteristics of low heat input, various intermetallic compounds would be produced in the welded joints. Abbasiet al. [47] found when welding Cu and Al using cold roll welding that although various intermetallic compounds such as Cu_3Al, Cu_4Al_3, CuAl and $CuAl_2$ existed in the welded joints, the growth rate of these intermetallic compounds was lower than that of friction stir welding. Laser welding has the characteristics of high energy density, fast welding speed and narrow heat-affected zone, which can further reduce the generation of intermetallic compounds. Moreover, when welding Cu and Al, filling silver, nickel, tin and other filler materials between the joints can also effectively reduce the formation of brittle phases [40–42].

Table 2. Summary of research conducted on laser beam welding of steel and aluminum.

No.	Materials	Optimum Laser Parameters	Main Outcomes	Intermetallics	Ref. (year)
1	1060 Al 316L stainless steel	Power: 285 W Speed: 4 mm·s^{-1}	The mechanical properties of the joint are related to the penetration depth	Not reported	[29] (2016)
2	5754 Al 301 stainless steel	Power: 2 kW Speed: 1.4 m·min	Applying magnetic field can reduce grain size and stabilize weld quality	Fe_2Al_5 $FeAl_3$	[30] (2016)
3	Low carbon steel st14 5754 Al	Power: 200 W Peak power: 1.43 kWSpeed: 5 mm/s	Increasing pulse time, pulse peak power and overlapping factor will result in the formation of more intermetallic compounds in the weld	Fe_2Al_5 $FeAl_3$ $FeAl_2$	[31] (2010)
4	201 stainless steel 5052 Al	Not reported	The addition of nickel interlayer helps to improve the metallurgical reaction of aluminum and iron, forming $Al_{0.9}Ni_{1.1}$ to improve the mechanical properties of the weld	Fe_2Al_5 $FeAl_3$ $Al_{0.9}Ni_{1.1}$	[32] (2012)
5	Press-hardened steel 5052 Al	Power: 1.2 kW Speed: 12 mm/s	Using laser offset welding can improve the mechanical properties of weld	Fe_2Al_5 Fe_4Al_{13}	[33] (2020)
6	310S stainless steel 6061 Al	Power: 2.4 kW Speed: 1.5 m/min	Stainless steel surface helps to improve weld quality with laser cleaning before welding	Not reported	[34] (2021)
7	Hilumin steel 1050 Al	Power: 600 W Speed: 60 mm/s	With the increase in laser swing amplitude, the depth of weld decreases linearly, and the severity of weld cracking decreases significantly	Fe_2Al_5 Fe_4Al_{13} $FeAl_2$	[35] (2022)
8	DC04 steel 6016-T4 Al	Power: 2 kW Speed: 1 m/min	The strength of the assemblies is shown to increase linearly with the reaction layer width	Not reported	[36] (2008)
9	H220YD 6061 Al	Power: 2600 W Speed: 1 m/min	By laser filling wire welding, the fused aluminum alloy and the filling wire can be brazed to galvanized solid steel	Al_8Fe_2Si $Al_{13}Fe_4$ Al_2Fe	[37] (2013)
10	DP590 6061-T6 Al	Power: 2 kW Speed: 0.5 m/min	The joint produced with the AlSi5 filler metal had the highest tensile strength and largest fracture displacement.	Fe_2Al_5 $FeAl_3$ Fe_2Al_8Si $Fe(Al,Si)_3$	[38] (2018)

Table 3. Properties of important intermetallics between Al and Cu.

Phase	Cu Content (at.%)	Structure	Microhardness (HV)	Density (g/cm^3)	Specific Resistance (μΩ cm)
$CuAl_2$	33	Body-centered tetragonal	630	4.34	8
CuAl	51	Body-centered orthorhombic	905	5.13	11.4
Cu_4Al_3	55.5	Monoclinic	930	NA	12.2
Cu_9Al_4	66	Body-centered cubic	770	6.43	14.2

Ali et al. [49] used 1050Al 0.75 mm thick and coated 70 μm nickel (Ni) thin layer of 99.5 mm thick and 1.5 mm thick AA40 aluminum alloy copper for lap welding. They found that the weld width increased with the increase in laser power and decreased with the increase in welding speed. With the increase in heat input, the depth of the weld deepened continuously towards the same side, and there were defects such as cracks and pores in the weld. Under high temperature conditions, a large number of Al–Cu eutectic alloys (α-Al + Al_2Cu) existed in the form of dendrites in the fusion zone of the weld, as shown in Figure 10. The high temperature also made the highly brittle Al_4Cu_9 phase distributed in the weld, making the brittleness of the weld become higher. The heat input also significantly affected the contact resistance of the weld.

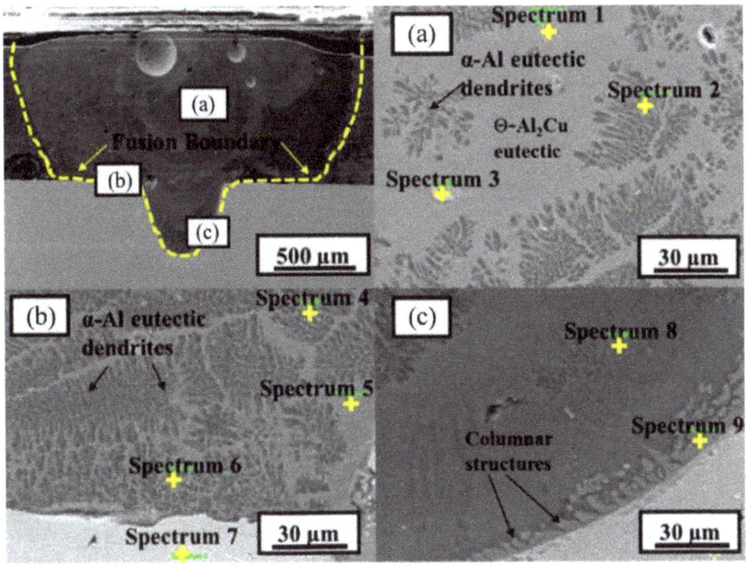

Figure 10. SEM images (power = 1500 W, welding speed = 30 mm/s); (**a**–**c**) are the magnified SEM images of microstructures at location **a**, **b** and **c** in weld fusion zone, respectively [49].

Lee et al. [50] used Al and Cu with purity of 99.99% as electrode samples. Al and Cu were used as the upper part of the lap joint for comparative test. A large amount of $CuAl_2$ was formed in the weld, and obvious α(Cu) α phase, $CuAl_2$ phase and Al + $CuAl_2$ phase could be seen, as shown in Figure 11b,c. When welding with Al as the upper material, Cu was evenly distributed throughout the area except for the area where molten Cu penetrates the Al melt. When Cu was used as the upper material, the Cu mixing zone was distributed along the lower Al layer, because Cu mixing was heavier than molten Al, so it diffuses downward, as shown in Figure 11a. Further analyzing the effect of welding speed on welding quality, they found that $CuAl_2$, Cu_9Al_4 and CuAl intermetallic compounds could be observed in the weld at a welding speed of 10 m/min. At a higher welding speed of 50 m/min, the formation of intermetallic compounds was inhibited. In addition, with the increase in welding speed, the tensile strength was increased, when the welding speed was 50 m/min, the tensile strength of the top aluminum reached 160 MPa, and the tensile strength of the bottom aluminum reached 205 MPa.

Hailat et al. [51] studied the continuous laser welding of 3003 aluminum and 110 copper. They welded two sets, one with tin as a sandwich and the other without tin. In welds with Sn interlayers, large pores could be seen in aluminum. However, the breaks occur far away from these pores, so they did not appear to affect joint strength, and the welds of tin-filled metals exhibit better bond shear strength, possibly due to the formation of Cu_6Sn_5 and Cu_3Sn.

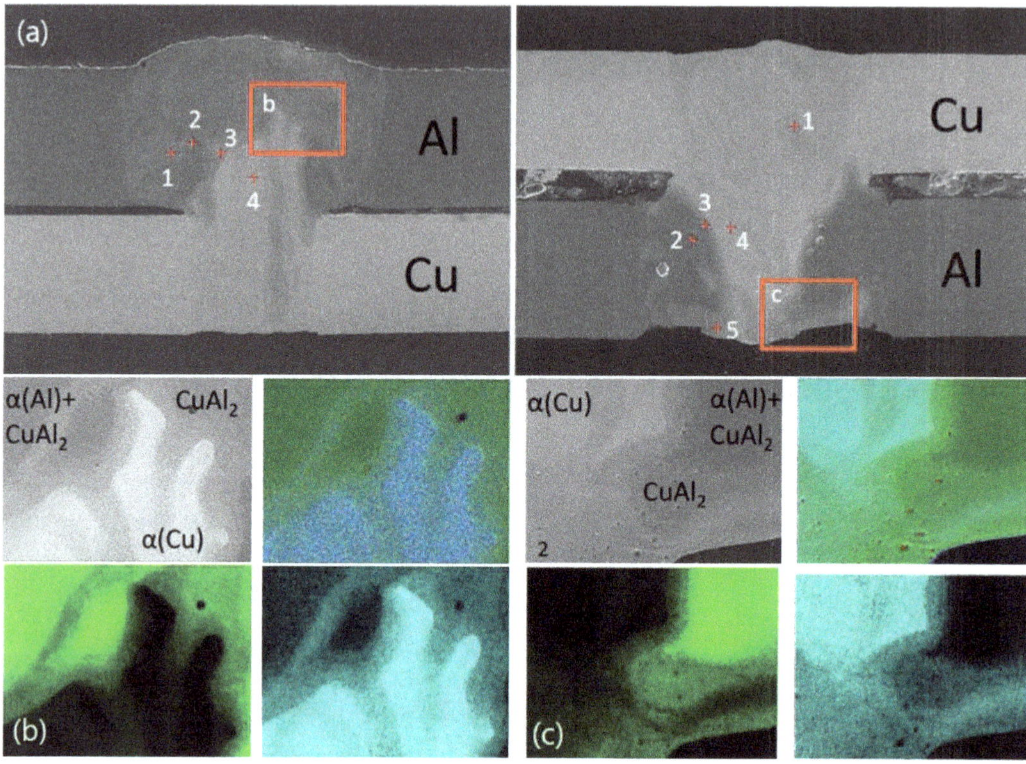

Figure 11. (**a**) Scanning electron microscopy images of the welded specimens. (**b**,**c**) EDS mapping images showing the distribution of Al and Cu to observe phase formation at a welding speed of 400 mm/s with: (**b**) Al as the upper layer and (**c**) Cu as the upper layer [50].

Xue et al. [52] observed the microstructure of the intermediate layer in laser welded copper aluminum lap joints. Tensile shear tests were conducted, and the fracture morphology was analyzed using SEM and EDS. The results indicate that there are several different regions in the interlayer of the weld seam, with different morphological and compositional characteristics, as shown in Figure 12. Banded and cellular structures were observed in the hypereutectic zone. The eutectic zone had a layered structure, the narrowest and thinnest. The thickest and widest dendritic structure was obtained in the hypoeutectic region. The joint fracture developed in the dendritic hypoeutectic zone, and the fracture mode was a combination of brittleness and shear. The maximum shear load of Cu–Al joints decreases with the increase in primary dendrite arm spacing and the growth of secondary dendrites in the hypoeutectic region, which was caused by an increase in laser power. Fusion welding of aluminum and copper, being dissimilar materials, is difficult, because brittle intermetallic compounds are formed in the welding zone, the weldability is poor, and the chemical, mechanical and thermal properties of welded joints are different. Due to the inevitable formation of brittle intermetallic compounds, the connection of aluminum and copper plates presents a metallurgical challenge. Therefore, it is necessary to effectively inhibit the formation and growth of Al–Cu intermetallic compounds. For the welding of dissimilar aluminum and copper sheets, there is no systematic work to reduce these defects.

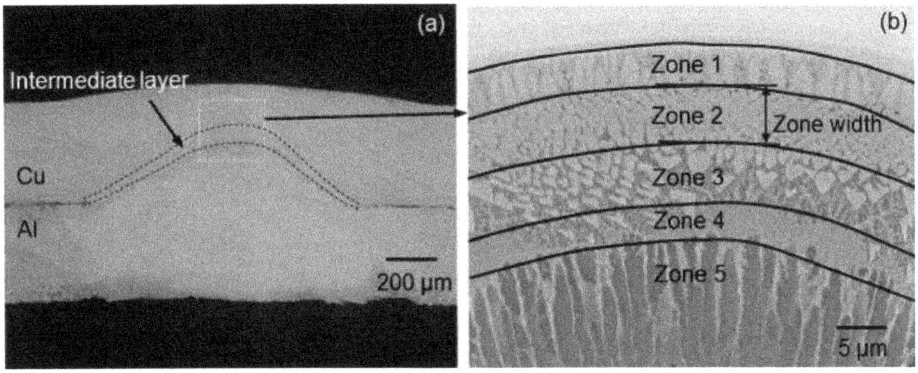

Figure 12. Micrograph of Cu–Al joint cross-section. (**a**) Optical micrograph; (**b**) SEM image of intermediate layer marked by the dashed square box in (**a**). Reprinted with permission from Ref. [52]. Copyright 2013, Elsevier.

Lee et al. [40] focused on the effect of welding speed on the quality of single-mode fiber laser lap welded joints of aluminum and copper sheets. They found that the intermetallic compounds are easy to form in a wide area at the welding speed of 1 kW and 10 m min^{-1}, while the intermetallic compounds were greatly reduced at the higher welding speed of 50 m min^{-1}. The width of the intermetallic compound decreased sharply to about 5 mm. According to the results of tensile shear test, the different loads and advantages of Al (upper) copper (lower) and copper (upper) process (lower) welding joints were almost equal to the welding speed of 50 m min^{-1} in most cases, and the loads of different welded joints were higher than those of similar devices, which were almost equivalent to the similar speed of Cu–Cu, and the tensile shear strength of the weld metal increases with the increase in the welding speed. Therefore, by inhibiting the formation of intermetallic compounds, solid laser welded joints could be produced at extremely high welding speed.

When welding Cu and AA4047 aluminum, Mai et al. [48] found that a weld without cracks could be obtained by shifting the laser to the aluminum side by 0.2 mm, and the weld had a higher hardness than the base material. Weigl et al. [53] used $AlSi_{12}$ as filler material for laser welding of pure copper and pure aluminum, respectively. They found that both $AlSi_{12}$ and $CuSi_3$ intermetallic compounds in the weld increased the ductility of the joint, and $AlSi_{12}$ with a higher Si content was more effective.

At present, the main research direction of welding between aluminum and copper has been the optimization of process parameters and the use of interlayer, and the subsequent research can add beam oscillation. Although the optimization of process parameters has yielded preliminary results, the potential formation of intermetallic compounds still needs further research. Table 4 summarizes the current research on aluminum and copper dissimilar laser welding.

2.3. Copper and Steel

In the welding of electric vehicle batteries, there are many types of welding between copper and steel, and Table 5 shows the room temperature properties of copper and iron. From the table, it can be seen that there are significant differences in the physical properties of copper and iron, especially the differences in melting temperature and thermal conductivity, making welding the two metals challenging [54]. In the Fe and Cu phase diagrams, there is a wide metastable miscibility gap at high temperatures [55]. In laser welding of steel and copper, liquid phase separation is a common feature due to the separation of undercooled Fe–Cu liquid into droplets of iron and copper [56]. Another major problem is that hot cracks appear in the welding zone or the heat-affected zone (HAZ) of the steel due to the penetration of Cu into the grain boundary [57].

Table 4. Summary of research conducted on laser beam welding of steel and aluminum.

No.	Materials	Optimum Laser Parameters	Main Outcomes	Intermetallics	Ref. (Year)
1	Cu99.5% AA 1050	Power: 1600 W Speed: 30 mm/s	The greater the heat input, the more intermetallic compounds are generated. The resistance decreases as the welding speed decreases.	Al_4Cu_9 Al_2Cu	[49] (2022)
2	Cu99.9% Al99.9%	Power: 2000 W Speed: 400 mm/s	The fracture after welding is mainly on the copper side.	$CuAl_2$	[50] (2022)
3	Cu 110-H00 Al 3003-H14	Power: 500 W Speed: 1 m/min	Adding tin alloy foil as interlayer can improve the mechanical properties of weld.	Not reported	[51] (2011)
4	T2 Cu 1060 Al	Power: 1450 W Speed: 100 mm/s	The microstructure of the subeutectic zone will greatly affect the shear resistance of the joint.	CuAl $CuAl_2$	[52] (2014)
5	Cu99.57% A1050 Al	Power: 1 Kw Speed: 10 m/min	Increasing the welding speed helps to reduce the content of intermetallic compounds in the weld.	CuAl $CuAl_2$	[40] (2013)
6	Oxygen-free Cu 4047 Al	Not reported	Controlling the melting ratio of metals is an important factor for defect-free welding of dissimilar metals.	Not reported	[48] (2004)
7	Pure Cu Pure Al	Not reported	The aluminum filler alloy AlSi12 produces a more uniform elemental mixture and a significantly enhanced ductility.	Not reported	[53] (2011)

Table 5. Summary of the room temperature properties of Al, Cu, Fe and Ni.

Metal	Melting Temperature (K)	Boiling Temperature (K)	Density ($Kg\ m^{-3}$)	Thermal Conductivity ($W\ m^{-1}\ K^{-1}$)	Thermal Expansion Coefficient ($10^6 K^{-1}$)
Fe	1809	3133	7870	78	12.1
Al	933	2739	2700	238	23.5
Cu	1356	2833	8930	398	17
Ni	1728	3188	8900	89	13.3

Joshi et al. [58] proposed the idea of using laser offset to weld copper and stainless steel butt joints. They shifted the laser to the stainless steel side, and there were obvious stainless steel particles in the weld after welding, and there were obvious solidification cracks after welding. Meng et al. [59], when using laser-arc composite welding to weld T2 copper/304 stainless steel, found that the welding position tilted to the pure copper side can effectively improve the welding quality. They found that when the welding position shifted to the pure copper side, the melting of the stainless steel gradually decreased, and the weld after welding had obvious depression on the surface and became full on the surface, as shown in Figure 13. They finally measured the tensile strength of the weld to 215 MPa, which was close to the tensile strength of T2 copper. Chen et al. [60], using laser welding to weld copper/stainless steel joints, also found that the laser offset to the stainless steel side could create a brazing effect, and they believed that the copper content in the joint should be limited to reduce the generation of cracks.

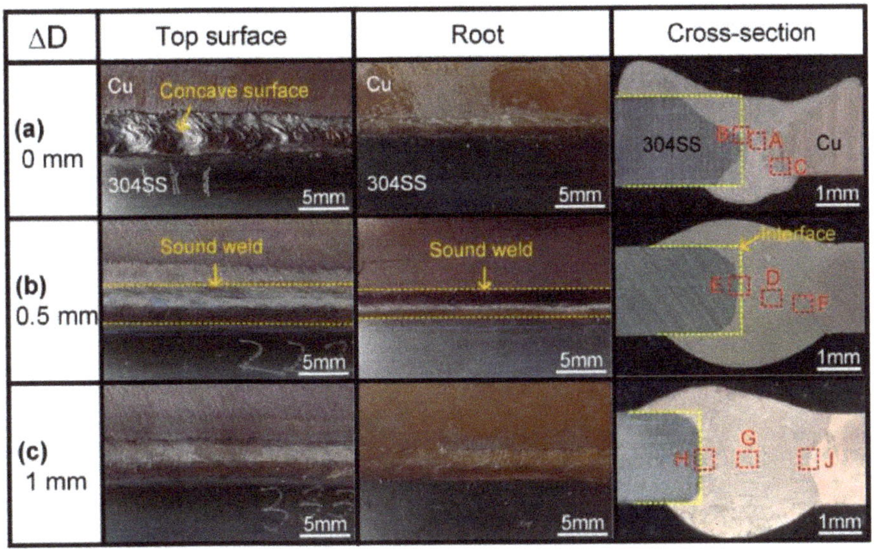

Figure 13. Surface and cross-sectional morphologies of the Cu/304SS dissimilar joints: ΔD = 0 mm, ΔD = 0.5 mm and ΔD = 1 mm. (**a**) Laser power = 3 kW, Arc current = 60 A, Welding speed = 1 m/min, Laser offset = 0 mm, (**b**) Laser power = 3 kW, Arc current = 60 A, Welding speed = 1 m/min, Laser offset = 0.5 mm, (**c**) Laser power = kW, Arc current = 60 A, Welding speed = 1 m/min, Laser offset = 1.0 mm. Reprinted with permission from Ref. [59]. Copyright 2019, Elsevier.

Fei et al. [61] conducted laser welding experiments on pure copper and stainless steel under the condition of annular beam oscillation. It was found that the crack resistance of weld could be improved effectively by adding oscillation. As could be seen from Figure 14, after adding oscillation, the microstructure in Figure 14d was significantly finer than that in Figure 14a, and the grain microstructure at the grain boundaries also presented discrete distribution, and the grain direction changed from single direction to multi-direction. When ring oscillating laser welding was used, the structure changed from blocky structure to spherical structure compared with Figure 14e, and the ($\alpha + \varepsilon$) structure was also evenly distributed in the welding area. The addition of beam shaking promotes solute migration and enhanced the consistency of plastic deformation in grains. Moreover, due to the larger range of action in the solidification process, the energy input and temperature field distribution were changed, and the thermal stress in the weld was also reduced. Under the action of beam vibration oscillation, Fe diffused into the Cu matrix, forming blocky ($\alpha + \varepsilon$) and spherical ($\alpha + \mu$) structures, as shown in Figure 14f. In addition, due to the driving action, the dissolution amount of copper in the weld increased, and the microstructure gradient of the weld decreased.

The microstructure with the addition of beam oscillation, as shown in Figure 15, was significantly refined compared with that without beam oscillation, as shown in Figure 15a. During the solidification process, the increase in the number of grain boundaries significantly reduced the stress concentration at the grain boundaries and improved the deformation. After adding the beam vibration, because the microstructure was more refined, the composition distribution was more uniform, and the strength and shape of the joint were effectively improved. The tensile strength and elongation of the joint were 282.82 MPa and 2.68 mm, respectively. Compared with the joint without beam vibration, the strength was increased by 10.41% and the tensile lift rate was increased by 53.14%, as shown in Figure 16.

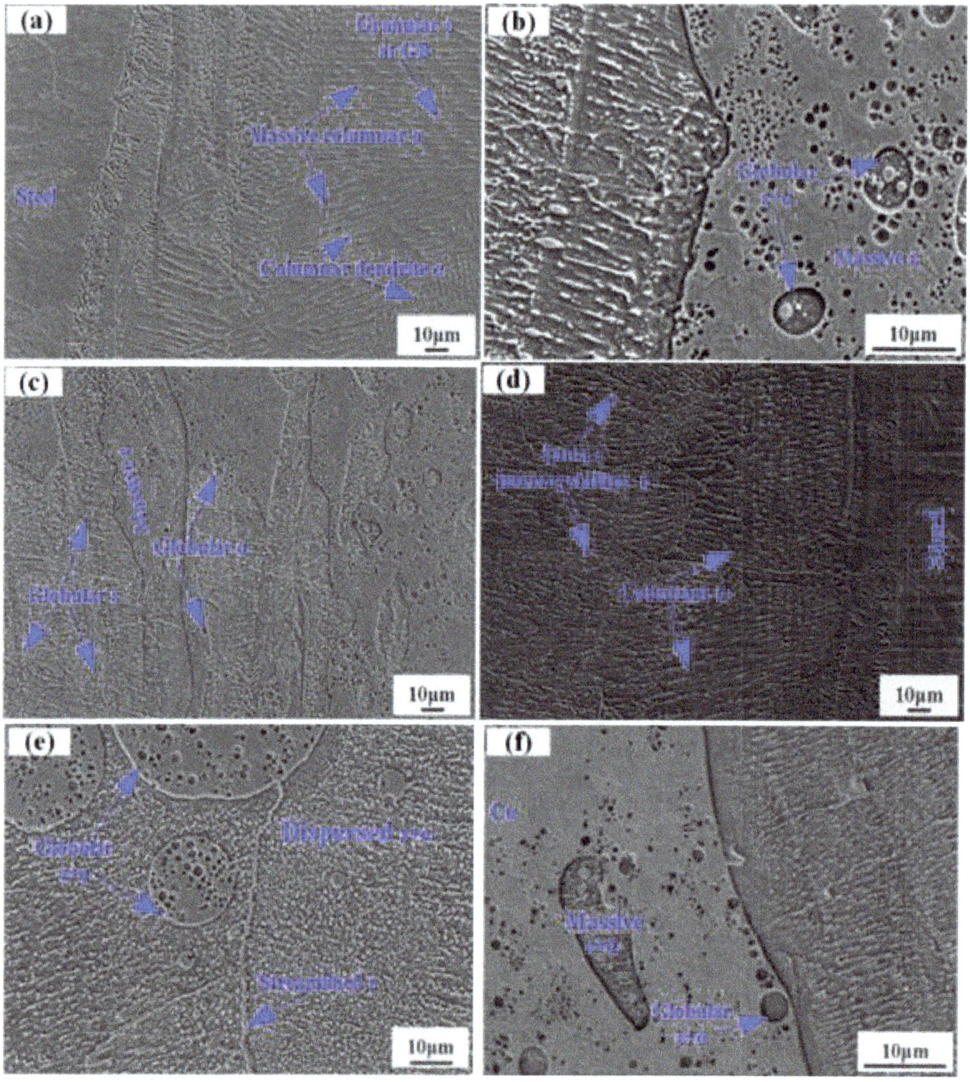

Figure 14. Microstructure of joints welded in the shortage of beam oscillation (**a–c**) and with beam oscillation (0.5 mm, 250 Hz) applied (**d–f**): (**a**/**d**) joint microstructures near the steel matrix; (**b**/**e**) joint microstructures in the middle weld; and (**c**/**f**) joint microstructures near the Cu matrix. Reprinted with permission from Ref. [61]. Copyright 2022, Elsevier.

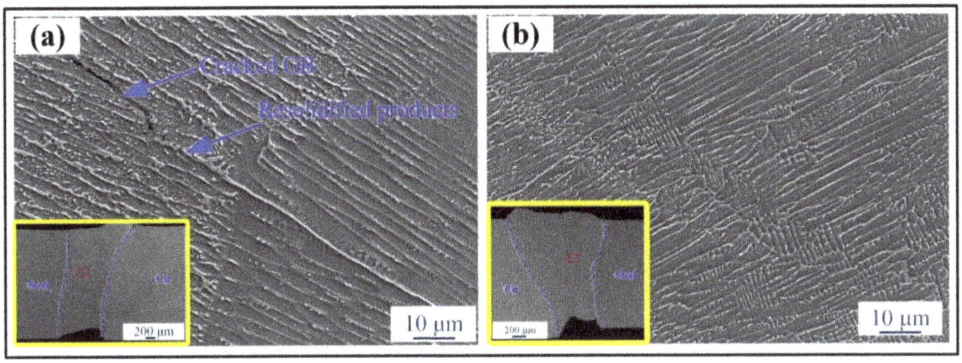

Figure 15. SEM micrographs of weld microstructures: (**a**) weld microstructures without exposure to beam oscillation and (**b**) weld microstructures at 0.5 mm, 250 Hz. Reprinted with permission from Ref. [61]. Copyright 2022, Elsevier.

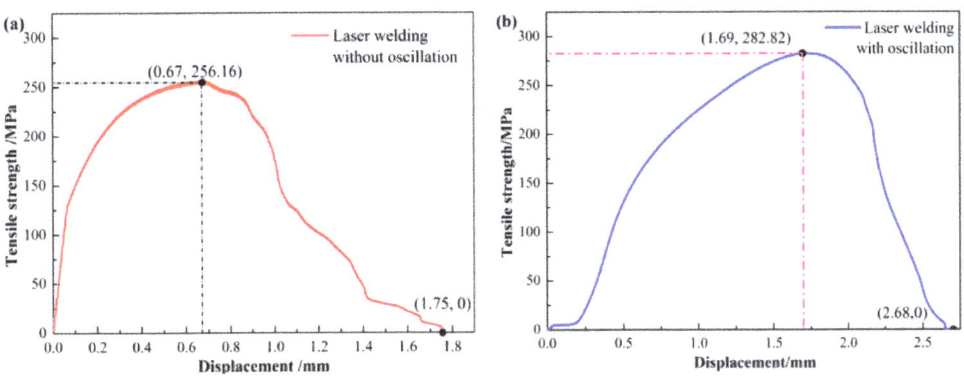

Figure 16. Tensile strength of joints welded: (**a**) with beam oscillation and (**b**) without beam oscillation. Reprinted with permission from Ref. [61]. Copyright 2022, Elsevier.

Mannucci et al. [62] welded 316L austenitic stainless steel with copper and found that the tensile property of the weld was mainly determined by the formation of heat-affected zone in solid copper. The melting zone of copper and stainless steel had a strongly asymmetrical shape, as shown in Figure 17, and the stainless part of the weld was much more developed and had an hourglass shape due to two eddies created by Marangoni convection. No heat-affected zone (HAZ) was formed in 316L side. The copper side of the weld was almost straight and bordered by a wide HAZ. Under the laser power, the Cu content in the melting zone increases with an increase in the laser power but never reached 50%, because Cu has a very high thermal diffusivity compared with 316L. Only the welded joint produced by the laser offset to the copper side had good performance.

Li et al. [63] found that the defects of laser welding between stainless steel and copper were mainly due to liquefaction cracking in the heat-affected zone of stainless steel and porosity in the fusion zone. Due to the presence of Fe–Cu compounds at the grain boundaries of the heat-affected zone, the bonding force between grains was affected and the sensitivity of weld cracks was increased. They found that the formation of liquefaction cracks could be divided into three stages, as shown in Figure 18. In the first stage, copper atoms permeated continuously along the grain boundaries, and cracks began to spread at the grain boundaries. In the second stage, Fe–Cu compounds accumulated at the grain boundaries, which led to the destruction of the bonding strength between grains and the initiation of cracks. In the third stage, with the increase in heat input in the laser welding process, the thermal stress increased significantly, causing the small crack at the grain boundary to expand into a large crack. The length of the crack increased with the increase in the heat input, as shown in Figure 19. However, when the heat input reached 125 KJ/m, the crack length began to decrease. This was because the molten copper present in the crack has self-healing properties due to a further increase in temperature. Although this self-healing property exists, the heat input should also be reduced during welding to better control the welding quality. In addition, they found that the porosity was largely due to the instability of the keyhole during the welding process, independent of the liquefaction cracking of the heat-affected zone. The laser focus during welding could be shifted to the stainless steel side, which could change the flow of the liquid metal and increase the stirring effect, which could help eliminate pores.

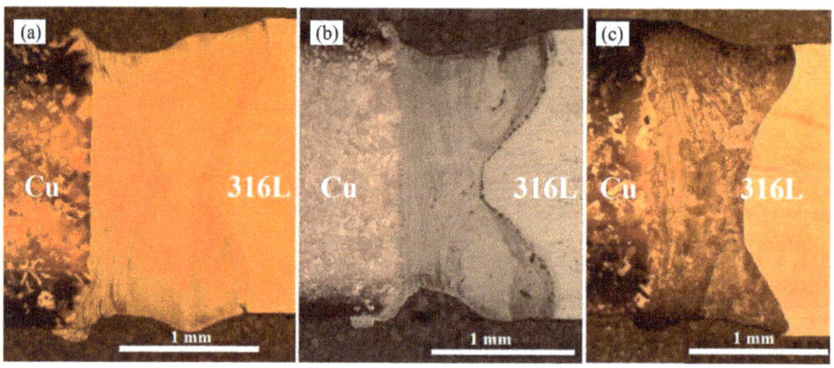

Figure 17. Typical weld cross-sections with beam shift: (**a**) 0.4 mm on steel (**b**) zero and 0.2 mm on copper and (**c**) 0.2 mm on copper [62].

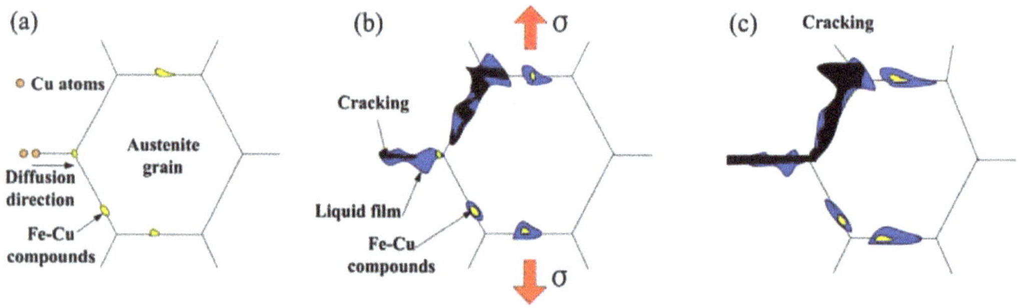

Figure 18. Liquation cracking model: (**a**) incubation, (**b**) initiation and (**c**) growth. Reprinted with permission from Ref. [63]. Copyright 2020, Elsevier.

Shen et al. [64] found that during laser welding of 316 stainless steel and oxygen-free copper, after the laser focus was shifted 0.4 mm to the copper side, no solidification cracks were detected in the weld after welding, and the copper content in the weld was 80%. Shaikh et al. [15] found in laser welding of copper and steel containing nickel coatings that the laser power, pulse opening time and frequency were positively correlated with the shear strength of the joint. And as the mixing degree of steel and aluminum increased, the strength of the joint increased.

Welding of butt joints of copper and steel has been studied. However, further studies of the microstructure of the weld zone and the interaction between the two materials are needed. Lap joints suitable for battery welding need to be paid more attention, and a sandwich can be added when welding. Table 6 summarizes the studies conducted on different laser welding of steel and copper.

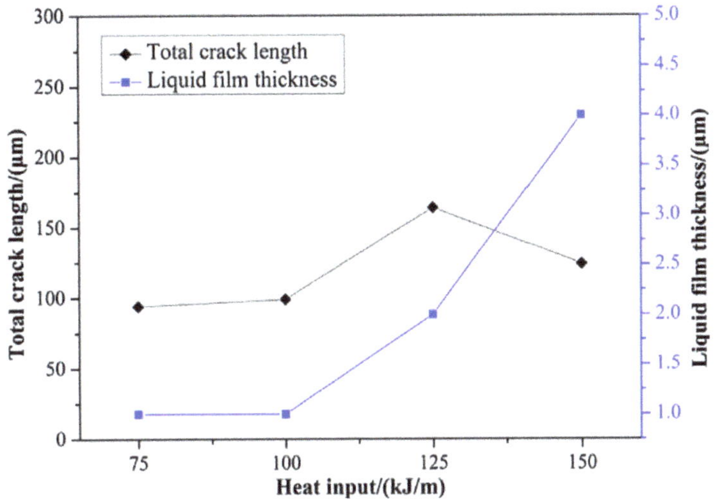

Figure 19. The correlation between the total crack length and heat input. Reprinted with permission from Ref. [63]. Copyright 2020, Elsevier.

Table 6. Summary of research conducted on laser beam welding of steel and copper.

No.	Materials	Optimum Laser Parameters	Main Outcomes	Intermetallics	Ref. (Year)
1	304L stainless steel ETP Cu	Power: 1000 W Speed: 300 mm/min	A defect-free bimetallic joint between Cu and SS can be obtained by laser beam welding technique with fusion welding mode.	Not reported	[58] (2019)
2	201 stainless steel T2 Cu	Not reported	Welding of copper/steel when welding—brazing and fusion welding depends on the welding parameters.	Not reported	[60] (2013)
3	304 stainless steel T2 Cu	Power: 4 kW Speed: 3 m/min	Adding beam oscillation in laser welding can refine grain and enhance weld mechanical properties.	Not reported	[61] (2022)
4	316L stainless steel CW Cu	Not reported	Excessive laser power can lead to the formation of thermal cracks.	Not reported	[62] (2018
5	304 stainless steel T2 Cu	Not reported	The susceptibility to HAZ liquation cracking can be effectively lowered by controlling the heat input during laser welding.	$Cu_{40}Fe_{60}$ Cu_xFe_{1-x}	[63] (2020)
6	316 stainless steel oxygen-free Cu	Not reported	By adjusting welding parameters, weld defects such as shrinkage, porosity, solidification cracks, etc., can be eliminated.	Not reported	[64] (2004)
7	Steel-Hilumin Cu	Power: 60 W Speed: 500 mm/min	Mechanical strength of the joint is highly correlated with electrical resistance and corresponding temperature rise at the joint.	Not reported	[15] (2019)

3. Welding of Battery Housing

The sealing of the battery housing, especially for hard cases, requires a high-quality weld, and since the electrical components in the case are very sensitive to heat input, this requires a low heat input [65]. Shown in Figure 20 is a Tesla 4680 battery. The top and bottom of the battery case need to be welded. Welding is required to have a stable weld depth, and there can be no molten drop in the tank because the molten drop will destroy the very sensitive electrolyte. After welding, in addition to assuring that the weld has good mechanical properties, it is crucial to ensure that the battery shell cannot be deformed. Therefore, laser welding has the characteristics of fast speed, small heat input and small range of action, which is an ideal processing mode for this production step in battery production. The battery housing is mainly made of aluminum alloy and steel, so it is necessary to study the same metal welding of welding aluminum alloy and stainless steel separately.

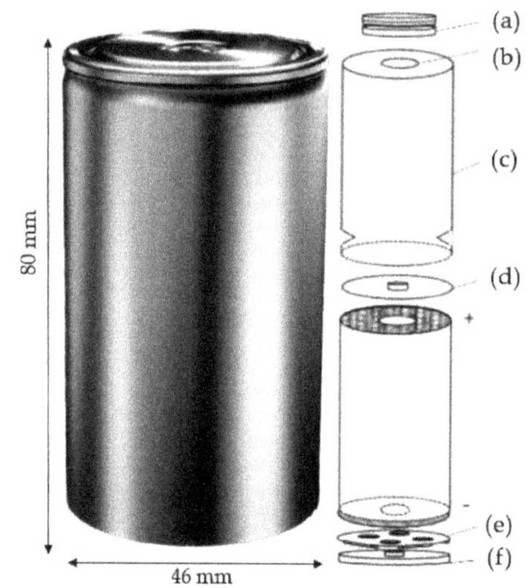

Figure 20. Tesla 4680 battery: (**a**) Pole cylinder, (**b**) front opening, (**c**) can, (**d**) positive collector plate, (**e**) negative collector plate and (**f**) cover plate.

3.1. Welding of Aluminum Battery Housing

Aluminum alloy has been widely used in various fields because of its advantages of light weight, high strength, good corrosion resistance and good formability such as in automobile manufacturing, aerospace, ship manufacturing, track manufacturing and so on. However, the welding of aluminum alloy has been facing various challenges because aluminum alloy has good thermal conductivity and high coefficient of thermal expansion, making it difficult to weld. Aluminum alloy welding is usually accompanied by cracks and pores and other welding defects [66–69]. The oxide film (Al_2O_3) and other organic impurities on the surface of aluminum alloy are easy to decompose at high temperature, which improves the porosity sensitivity of the weld. Gas phase components such as hydrogen are more soluble in the melt pool at high temperatures, and the faster cooling rate makes it difficult for them to escape and form pores [44].

In order to solve the problems of forming porosity and poor welding stability in the welding process of aluminum alloy, Wu et al. [70] proposed a new welding method. They used fiber laser with focused rotation and vertical oscillation to weld 1060 aluminum alloy. This method combined a vertical oscillation rotation of the focus along the direction of the beam. They found that when only vertical oscillation laser focusing welding was added, the width of the weld was relatively small, and with the increase in the oscillation amplitude, the shape of the weld surface first continuously improved, and when the oscillation amplitude continued to increase, the surface morphology of the weld began to gradually deteriorate. When the vertical oscillation amplitude was 2 mm, the weld surface formation was best, as shown in Figure 21. When the rotation radius was within the range of 0~0.45 mm, the formation of the weld surface gradually improved with the increase in the oscillation amplitude, and the surface formed had fish-scale ripples, which were dense, uniform and smooth. This indicated that focal rotation could improve the quality of weld surface formation. Further observation of the depth of the weld showed that, as shown in Figure 22, when the rotation radius was less than 0.15 mm, it had little influence on the penetration depth of the welds using vertical oscillation laser welding and has little change compared with no addition. The morphology of the weld section was thin and long nail shape. When the rotation radius of the laser was 0.45 mm, the weld section was shallow but wide nail shape, and the width of the weld and the middle part are relatively large. The weld morphology of focus rotating laser and vertical oscillation laser was similar to that of focus rotating laser welding only. It could be said that the depth of the weld was mainly affected by the amplitude of the vertical oscillation laser, and the weld appearance was mainly affected by the rotation radius of the laser focus rotation. They also found that adding the rotation of the focus during the welding process could effectively reduce the formation of porosity, and the larger the rotation radius, the lower was the porosity. When the rotation radius was 0~0.45 mm, the porosity decreased with the increase in the rotation radius. When the rotation radius was 0.45 mm, the porosity of the weld was reduced by 91% compared with the non-focused rotation and non-vertical oscillation, as shown in Figure 23.

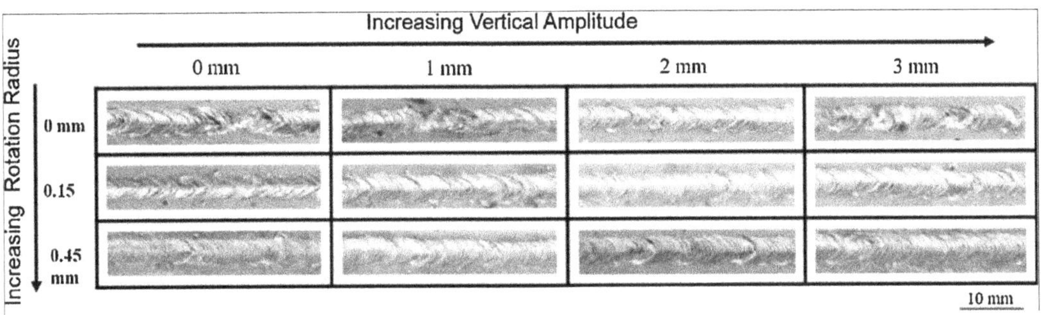

Figure 21. Surface morphologies of laser beam focus rotation and vertical oscillation welding ($f_{rot} = f_{osc} = 100$ Hz). Reprinted with permission from Ref. [70]. Copyright 2018, Elsevier.

Mauritz et al. [71] also adjusted the laser. On the basis of adding beam vibration, they applied the laser spot beam shaping technology to divide individual laser spots into four single beams with equal power, and then interacted with the material in a processing area, as shown in Figure 24. The dynamic situation of the molten pool was observed by high-speed photography, as shown in Figure 25. They found that the keyhole size of the molten pool significantly increased when the multi-focus technology was adopted. Compared with the average area of 0.13 mm^2 during single-beam welding, the average area of the multi-focus technology increased by 10.6 times, reaching 1.51 mm^2. Further dynamic observation showed that the stability of metal vapor in the molten pool is increased with the increase in keyhole area. Compared with single point welding, the keyhole of multi-focus method is constantly opened. Moreover, the standard deviation of keyhole size is effectively reduced by the multi-focal technique, and the area fluctuation of keyhole is reduced by 7.1% from 54.3% in single point welding. Compared with the single-point process, the multi-focus method doubles the length of the weld pool, which also allows the energy input to be more evenly distributed in space, effectively reducing the splash during welding. X-ray observation was performed on the section after welding, as shown in Figure 26. Compared with single-point welding, the size of pores in the weld was significantly reduced when the multi-focus technology was adopted for welding, so the new technology greatly improved the quality of the weld.

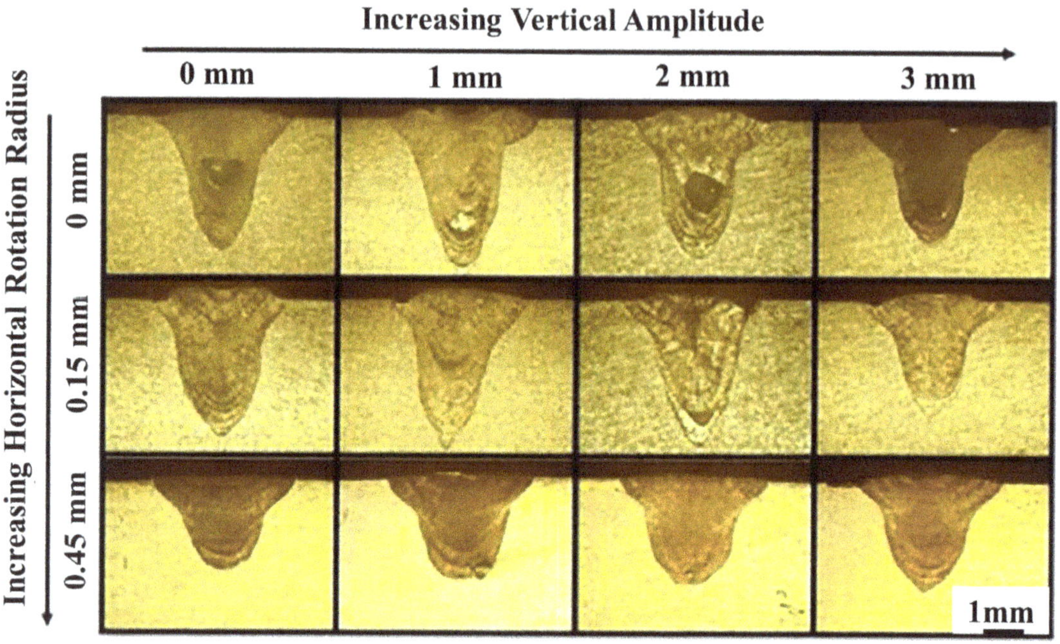

Figure 22. Effect of rotation radius and vertical oscillation amplitude on weld cross-section morphology ($f_{rot} = f_{osc}$ = 100 Hz). Reprinted with permission from Ref. [70]. Copyright 2018, Elsevier.

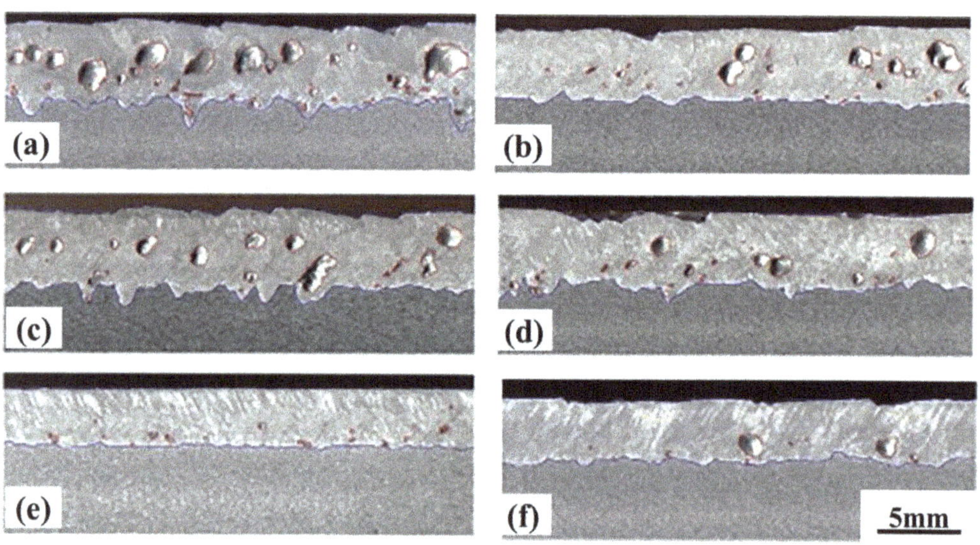

Figure 23. Pores in longitudinal section of welds under different rotation radius and vertical amplitude: (**a**) No oscillation, (**b**) A_{osc} = 2 mm, (**c**) R_{rot} = 0.15 mm, (**d**) R_{rot} = 0.15 mm and A_{osc} = 2 mm, (**e**) R_{rot} = 0.45 mm and (**f**) R_{rot} = 0.45 mm and A_{osc} = 2 mm. Reprinted with permission from Ref. [70]. Copyright 2018, Elsevier.

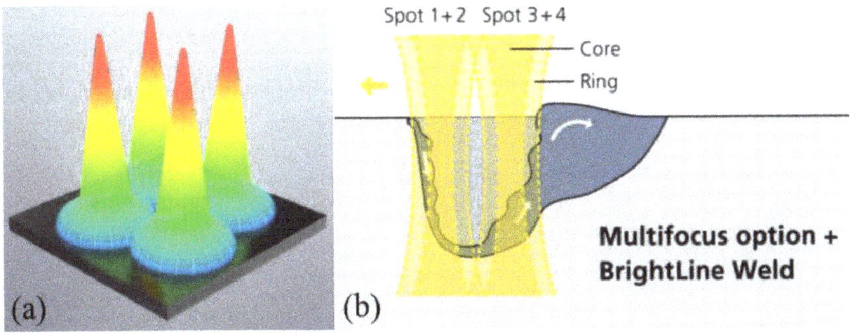

Figure 24. (**a**) Schematic diagram of laser power distribution of four spots formed by multi-focusing. (**b**) Schematic diagram of laser spot acting on molten pool [71].

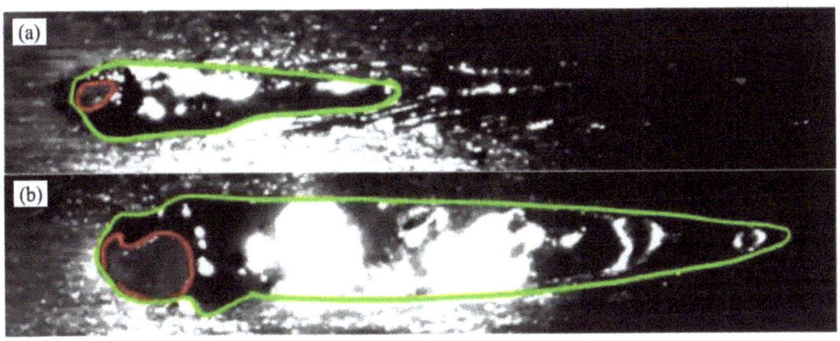

Figure 25. (**a**) Single spot welding with beam shaping. (**b**) Multi-focus approach with stabilized keyhole. [red: current keyhole size; green: current weld pool size] [71].

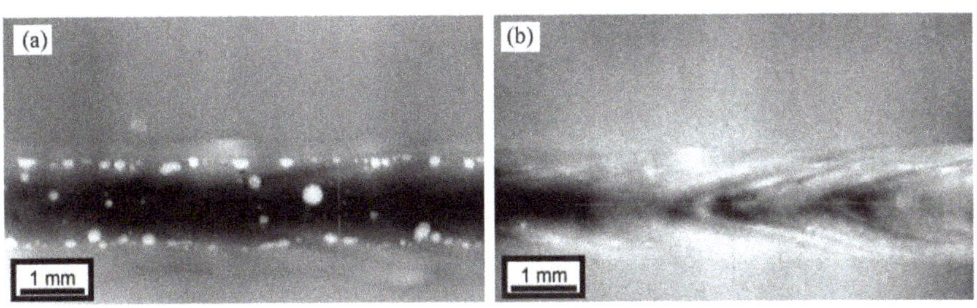

Figure 26. (**a**) X-ray image of single spot welding with beam shaping shows significant porosity in the weld seam. (**b**) X-ray image of multi-focus welding shows substantial reduction of the porosity in the weld seam [71].

Prieto et al. [72] shaped the laser spot and formed the laser into six shapes as shown in Figure 27 for welding experiments on AA3003. By comparison, they found that when the laser power was between 2.4~3.6 kW, the welding speed was 18 m/min, and the shape frequency was up to 1 MHz, and the best weld formation could be obtained by using spiral oscillation and infinite oscillation, as shown in Figure 28. Further analysis of the influence of welding speed on the welding form, they found that when the welding speed of spiral spot was increased from 6 m/min to 18 m/min, the penetration depth of the weld was reduced from 730 μm to 570 μm, and the change remained in the range of 20%. Similar results were obtained for welding with infinite spot shape. And by changing the graphic frequency from 1 kHz to 1 MHz, the weld width was basically constant. At the interface, the width of the weld began to decrease significantly only when the frequency was increased to 111 kHz, and the penetration depth began to increase significantly when the frequency was lower than 11 kHz. Good weld formation could be obtained after welding regardless of whether spiral or infinite light spots were used.

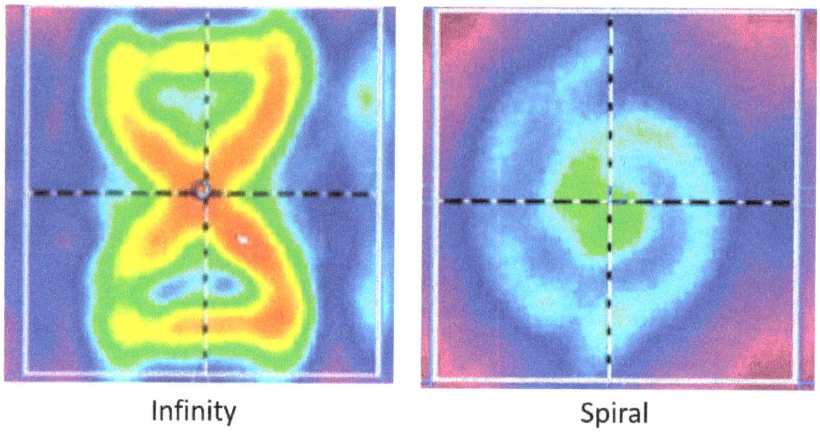

Figure 27. Diagram of light spot shape [72].

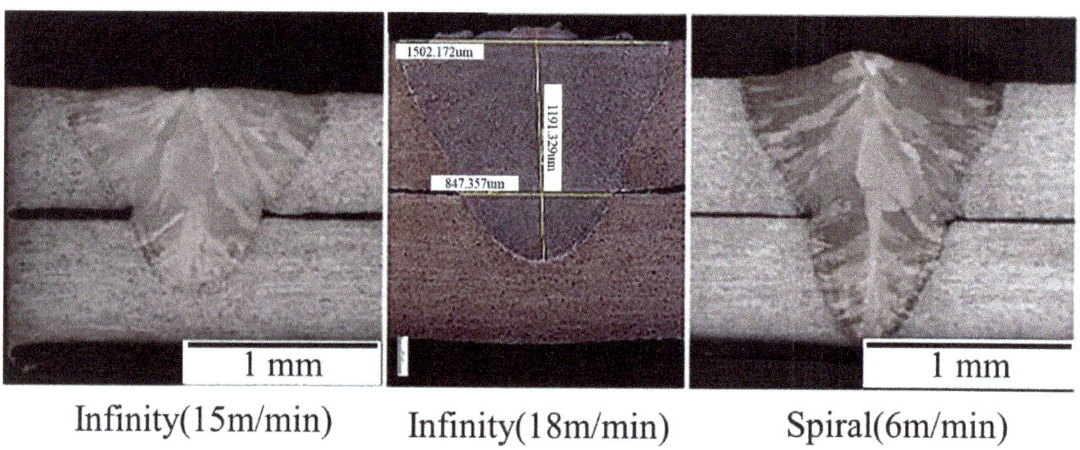

Figure 28. Weld forming under different spot shape and speed [72].

3.2. Welding of Steel Battery Housing

Stainless steel has been widely used in various applications because of its excellent mechanical properties, corrosion resistance and good welding performance, such as biomedical equipment, nuclear industry components, precision instruments, aviation, automotive industry, batteries and so on [73–78]. The welding of stainless steel usually does not require pre-welding or post-welding heat treatment, because unlike low-carbon steel alloy, there is no Martensite structure in HAZ [79]. Hot cracks are easy to occur during welding of austenitic stainless steel, especially for stainless steel without ferrite [80]. At present, it has been found that the hot crack in stainless steel welding is mainly related to the alloy composition and the content of impurities [81]. Sulfur and phosphorous impurities, in particular, expand the solidification range by forming eutectic liquids with low melting point [80,82,83]. The Hammar–Svensson chrome–nickel equivalent ratio (Creq/Nieq) during solidification also affects the crack sensitivity [84–86].

Yan et al. [87] used CO_2 laser to conduct welding experiments on 5 mm thick 304L stainless steel and successfully obtained good weld joints. They found that no significant cracks or voids were found in the welded joint. The width of the connector was 3.5 mm, and the fusion area of the connector was 6.7 mm^2. It could be seen from the microstructure of the joint that the columnar dendrites extended from the fusion boundary to the weld center line, and there was no obvious transition zone and heat-affected zone near the joint. The joint is composed of dark δ-Fe dendrites in austenite matrix, in which δ-Fe existed in a network structure in the fusion zone, and the dendrite spacing was 2–5 μm.

Danny et al. [88] used lap joint welding for stainless steel foil, and they used 60 μm 304 stainless steel foil. They gradually increased the power of the laser from 200 mJ to 400 mJ and the welding speed increased from 0.5 mm/s to 1.5 mm/s. After welding, they found that most of the lap welds showed good surface finish, without the common edge bites and bumps, but with the increase in welding speed and laser pulse energy, the weld roughness increased, accompanied by porosity and collapse. The optimal weld was obtained when the laser pulse energy was 250 mJ, the speed was 0.5 mm/s and the laser pulse frequency was 10 Hz, as shown in Figure 29. Through observation, it could be found that the secondary dendrite distance was submicron level due to the fast cooling speed of laser welding. Further observation by electron microscope showed that the austenite microstructure in the weld had high density dislocation, and there were carbide precipitates in some areas, as shown in Figure 30.

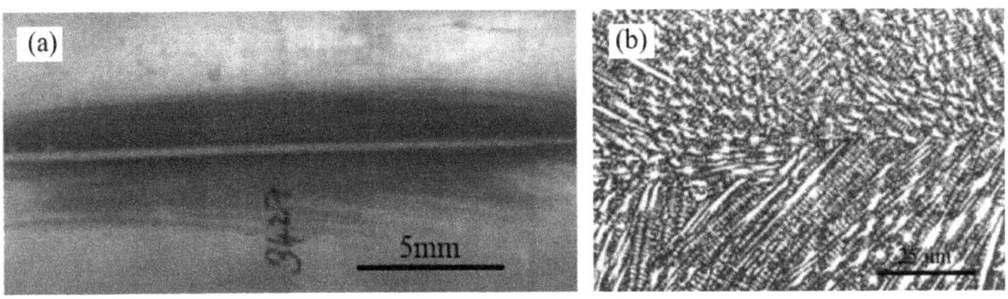

Figure 29. Macro morphology of weld (**a**) and microstructure of weld (**b**) (etchant: 10% oxalic acid). Reprinted with permission from Ref. [88]. Copyright 2008, Elsevier.

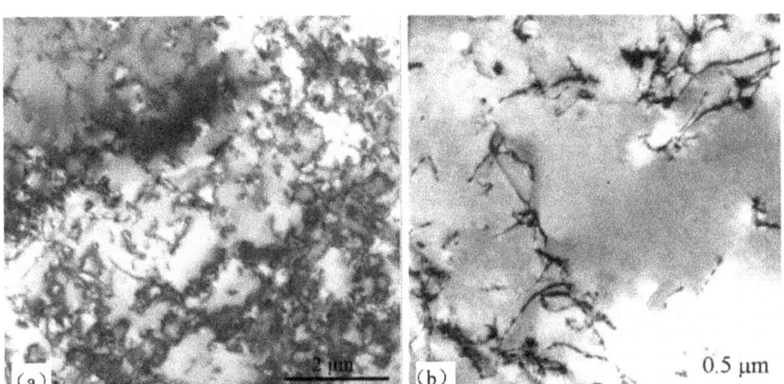

Figure 30. Transmission electron micrograph of LBW weld (**a**). Transmission electron micrograph of LBW weld showing the carbide preferentially precipitated at the dislocations (**b**). Reprinted with permission from Ref. [88]. Copyright 2008, Elsevier.

Zhang et al. [89] conducted deep penetration laser welding of 10 mm thick plate stainless steel and studied the influence of welding parameters on weld-formation. When the welding speed increased, the surface shape of the welds gradually became better, the width of the weld gradually became narrower from the width, and the weld nodules at the bottom of the weld also gradually became smaller. Only changing the defocus amount of laser, when the defocus amount was −10 mm, the weld joint forming best. By changing the types of protective gases used in welding process, they found that the maximum weld depth could be obtained by using helium as the protective gas, followed by nitrogen and argon as the shallowest weld penetration. When the laser power could be completely penetrated, no matter what kind of shielding gas was used, the appearance of the weld would not be affected. However, argon gas could be used as shielding gas to protect the bottom of the molten pool, as shown in Figure 31. And the most continuous smooth weld with no spatter and no porosity could be obtained. The weld was examined using XRD, as shown in Figure 32, and it could be seen that the composition of the weld was mainly composed of γ-Fe and δ-Fe, of which γ-Fe had a higher content than δ-F. By analyzing the joint of the weld with the optimal welding parameters, as shown in Figure 33a, it could be found that there is no obvious transition zone and heat-affected zone (HAZ) in the weld. The microstructure of the fusion zone was mainly columnar dendrites, which are symmetrically distributed along the weld center. The microstructure of the fusion zone center is equiaxed, as shown in Figure 33b,c. In Figure 33d, the dark dendritic structure was δ-F, where the lighter material was γ-Fe.

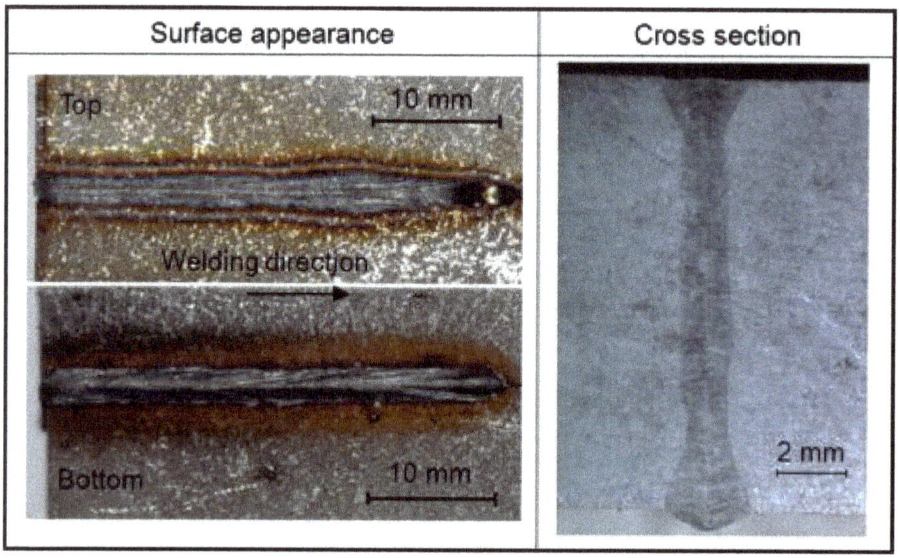

Figure 31. Surface appearance and cross-section of optimal butt joint welded at v = 2.4 m/min, Δ = −10 mm, q_{top} = 30 L/min (N_2) and bottom shielding gas flow (q_{bottom}) of 5 L/min (Ar). Reprinted with permission from Ref. [89]. Copyright 2014, Elsevier.

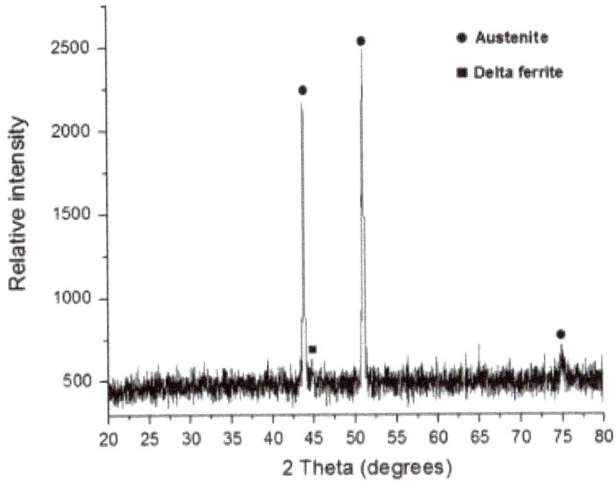

Figure 32. XRD patterns of the joint. Reprinted with permission from Ref. [89]. Copyright 2014, Elsevier.

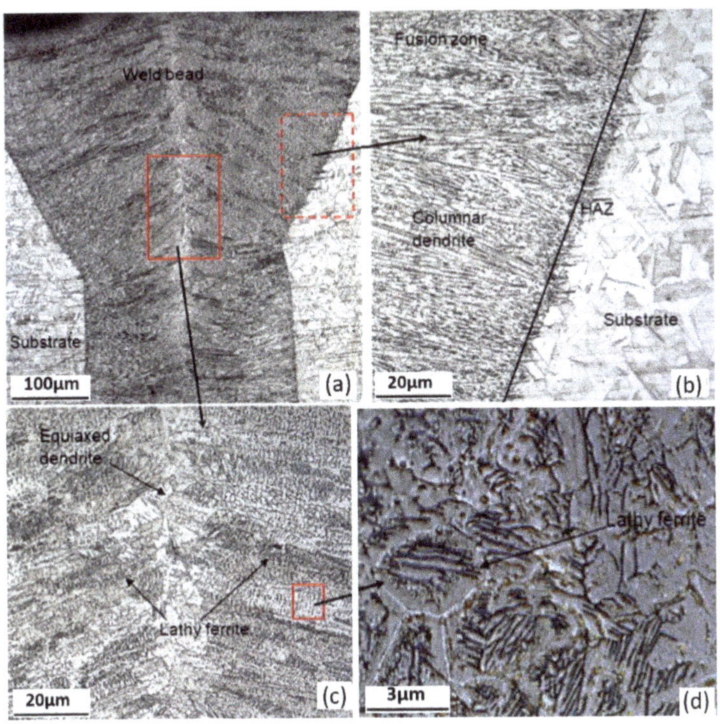

Figure 33. Microstructure of optimal butt joint: (**a**) optical micrograph of the entire joint, (**b**) enlarged micrograph between the substrate and fusion zone inside dashed rectangle in (**a**), (**c**) enlarged micrograph of the fusion center zone inside solid rectangle in (**a**), and (**d**) enlarged micrograph of the columnar dendrite zone inside solid square in (**c**). Reprinted with permission from Ref. [89]. Copyright 2014, Elsevier.

4. Summary and Outlook

Laser welding is a welding method with high energy density and non-contact and accurate heat input control, which can provide reliable weldability for the welding between dissimilar materials in the battery system of electric vehicles. The laser welding of dissimilar materials has made great progress in the past years. However, no matter the different laser light sources, the optimization of process parameters or the improvement of various joint structures are studied, and there will still be metallurgical defects such as incomplete bonding, brittle metal interphase, corrosion, excessive porosity and cracking. These defects affect the electrical performance and safety of the entire EV battery system. Therefore, if laser welding technology wants to be widely used in the manufacturing of electric vehicle batteries, further research is needed. This paper reviews the research progress of laser welding technology of steel–copper, steel–aluminum, al–copper, Al aluminum and steel–steel composites is reviewed. Based on the current research, the following suggestions are proposed for future research in this field:

(1) Because the material thickness used in electric vehicle batteries is generally low, it is necessary to optimize the process parameters, accurately control the heat input and improve the welding quality by controlling the thickness of intermetallic compounds. For example, the matching of laser power and welding speed as well as the appropriate beam oscillation frequency.

(2) Appropriate interlayers or coatings are used to change the formation of metal components during welding, so as to adjust the microstructure, improve mechanical properties and reduce resistance.

(3) At present, lasers are still mainly at infrared wavelength (1064 nm), and metals such as copper and aluminum have higher reflectance to the light of the sub-band and have higher absorption rate for blue light (450 nm) and green light (515 nm). Blue light and green light laser can be used for welding experiments.

(4) So far, most of the mechanical properties analyses of laser welded dissimilar material joints have analyzed the static mechanical properties of the joints and few have researched the fatigue properties of the joints. Because electric vehicles are often accompanied by bumps on the road, the performance of these joints under cyclic loads is important. Therefore, it is necessary to carry out more studies on mechanical properties of joints.

Author Contributions: Conceptualization: J.F., methodology: J.F. and Z.L., investigation: J.F., P.Z., H.Y., H.S. and Q.L., formal analysis: J.F. and Z.L., writing—original draft: J.F., data curation: P.Z. and D.W., supervision: P.Z., writing—review and editing: P.Z., R.L. and Q.W. visualization: P.Z. and T.S., validation: P.Z., funding acquisition: P.Z. All authors have read and agreed to the published version of the manuscript.

Funding: This research was supported by Foundation of Natural Science Foundation of China (52075317 and 51905333), Shanghai Sailing Program (21YF1432300), China Postdoctoral Science Foundation under Grant 2022T150400 and Class III Peak Discipline of Shanghai—Materials Science and Engineering (High-Energy Beam Intelligent Processing and Green Manufacturing).

Institutional Review Board Statement: Not applicable.

Informed Consent Statement: Not applicable.

Data Availability Statement: Data sharing is not applicable to this article.

Conflicts of Interest: The authors declare no conflict of interest.

References

1. Li, Z.; Khajepour, A.; Song, J. A comprehensive review of the key technologies for pure electric vehicles. *Energy* **2019**, *182*, 824–839. [CrossRef]
2. Eskander, S.M.S.U.; Fankhauser, S. Reduction in greenhouse gas emissions from national climate legislation. *Nat. Clim. Chang.* **2020**, *10*, 750–756. [CrossRef]

3. Wimbadi, R.W.; Djalante, R. From decarbonization to low carbon development and transition: A systematic literature review of the conceptualization of moving toward net-zero carbon dioxide emission (1995–2019). *J. Clean. Prod.* **2020**, *256*, 120307. [CrossRef]
4. Sun, X.; Li, Z.; Wang, X.; Li, C. Technology development of electric vehicles: A review. *Energies* **2019**, *13*, 90. [CrossRef]
5. Faria, R.; Moura, P.; Delgado, J.; De Almeida, A.T. A sustainability assessment of electric vehicles as a personal mobility system. *Energy Convers. Manag.* **2012**, *61*, 19–30. [CrossRef]
6. Whittingham, M.S. History, evolution, and future status of energy storage. In *Proceedings of the IEEE*; Institute of Electrical and Electronics Engineers Inc.: Manhattan, NY, USA, 2012; pp. 1518–1534. [CrossRef]
7. Becker, J.; Nemeth, T.; Wegmann, R.; Sauer, D.U. Dimensioning and optimization of hybrid Li-ion battery systems for EVs. *World Electr. Veh. J.* **2018**, *9*, 19. [CrossRef]
8. Das, A.; Li, D.; Williams, D.; Greenwood, D. Joining technologies for automotive battery systems manufacturing. *World Electr. Veh. J.* **2018**, *9*, 22. [CrossRef]
9. Canis, W. Federal and State Incentives Heighten Consumer Interest in Electric Vehicles. Available online: www.carhistory4u.com (accessed on 19 June 2023).
10. Zwicker, M.F.R.; Moghadam, M.; Zhang, W.; Nielsen, C.V. Automotive battery pack manufacturing—A review of battery to tab joining. *J. Adv. Join. Process.* **2020**, *1*, 100017. [CrossRef]
11. Kumar, N.; Masters, I.; Das, A. In-depth evaluation of laser-welded similar and dissimilar material tab-to-busbar electrical interconnects for electric vehicle battery pack. *J. Manuf. Process.* **2021**, *70*, 78–96. [CrossRef]
12. Brand, M.J.; Schmidt, P.A.; Zaeh, M.F.; Jossen, A. Welding techniques for battery cells and resulting electrical contact resistances. *J. Energy Storage* **2015**, *1*, 7–14. [CrossRef]
13. Zhang, C.; Li, L. A coupled thermal-mechanical analysis of ultrasonic bonding mechanism. *Metall. Mater. Trans. B Process Metall. Mater. Process. Sci.* **2009**, *40*, 196–207. [CrossRef]
14. Lee, S.; Kim, T.H.; Hu, S.J.; Cai, W.W.; Li, J.; Abell, J.A. Characterization of Joint Quality in Ultrasonic Welding of Battery Tabs MSEC2012-7410. Available online: http://proceedings.asmedigitalcollection.asme.org/pdfaccess.ashx?url=/data/conferences/asmep/75447/ (accessed on 19 June 2023).
15. Institutul de Cercetări Pentru Industria Electrotehnică, Association for Promoting Electric Vehicles in Romania, IEEE Circuits and Systems Society, Institute of Electrical and Electronics Engineers, IEEE Power & Energy Society, and Romania. Ministerul Educației și Cercetării Științifice. In Proceedings of the 2019 Electric Vehicles International Conference (EV): Icpe Solar Park, Bucharest, Romania, 3–4 October 2019.
16. Luan, B.L.; Campbell, G.; Gauthier, M.; Liu, X.Y.; Davidson, I.; Nagata, J.; Lépinay, M.; Bernier, F.; Argue, S. Surface Modification and Fabrication of Li-Ion Battery Components for Plug-In Hybrid Electric Vehicle. *ECS Trans.* **2010**, *25*, 59–71. [CrossRef]
17. Schweier, M.; Heins, J.F.; Haubold, M.W.; Zaeh, M.F. Spatter formation in laser welding with beam oscillation. *Phys. Procedia* **2013**, *41*, 20–30. [CrossRef]
18. Bonenberger, P.R. *The First Snap-Fit Handbook: Creating and Managing Attachments for Plastics Parts*; Carl Hanser Verlag GmbH Co. KG: Munich, Germany, 2016.
19. Niu, P.; Li, R.; Fan, Z.; Cao, P.; Zheng, D.; Wang, M.; Deng, C.; Yuan, T. Inhibiting cracking and improving strength for additive manufactured AlxCoCrFeNi high entropy alloy via changing crystal structure from BCC-to-FCC. *Addit. Manuf.* **2023**, *71*, 103584. [CrossRef]
20. Springer, H.; Kostka, A.; Santos, J.F.D.; Raabe, D. Influence of intermetallic phases and Kirkendall-porosity on the mechanical properties of joints between steel and aluminium alloys. *Mater. Sci. Eng. A* **2011**, *528*, 4630–4642. [CrossRef]
21. Kouadri-David, A. Study of metallurgic and mechanical properties of laser welded heterogeneous joints between DP600 galvanised steel and aluminium 6082. *Mater. Des.* **2014**, *54*, 184–195. [CrossRef]
22. Ghosh, M.; Chatterjee, S.; Mishra, B. The effect of intermetallics on the strength properties of diffusion bonds formed between Ti-5.5Al-2.4V and 304 stainless steel. *Mater. Sci. Eng. A* **2003**, *363*, 268–274. [CrossRef]
23. Bozzi, S.; Helbert-Etter, A.L.; Baudin, T.; Criqui, B.; Kerbiguet, J.G. Intermetallic compounds in Al 6016/IF-steel friction stir spot welds. *Mater. Sci. Eng. A* **2010**, *527*, 4505–4509. [CrossRef]
24. Qiu, R.; Iwamoto, C.; Satonaka, S. The influence of reaction layer on the strength of aluminum/steel joint welded by resistance spot welding. *Mater. Charact.* **2009**, *60*, 156–159. [CrossRef]
25. Zhou, D.W.; Liu, J.S.; Lu, Y.Z.; Xu, S.H. Effect of adding powder on joint properties of laser penetration welding for dual phase steel and aluminum alloy. *Opt. Laser Technol.* **2017**, *94*, 171–179. [CrossRef]
26. Shahverdi, H.R.; Ghomashchi, M.R.; Shabestari, S.; Hejazi, J. Microstructural analysis of interfacial reaction between molten aluminium and solid iron. *J. Mater. Process. Technol.* **2002**, *124*, 345–352. [CrossRef]
27. Cinca, N.; Lima, C.R.C.; Guilemany, J.M. An overview of intermetallics research and application: Status of thermal spray coatings. *J. Mater. Res. Technol.* **2013**, *2*, 75–86. [CrossRef]
28. Indhu, R.; Soundarapandian, S.; Vijayaraghavan, L. Yb: YAG laser welding of dual phase steel to aluminium alloy. *J. Mater. Process. Technol.* **2018**, *262*, 411–421. [CrossRef]
29. Yang, J.; Li, Y.L.; Zhang, H. Microstructure and mechanical properties of pulsed laser welded Al/steel dissimilar joint. *Trans. Nonferrous Met. Soc. China* **2016**, *26*, 994–1002. [CrossRef]

30. Chen, R.; Wang, C.; Jiang, P.; Shao, X.; Zhao, Z.; Gao, Z.; Yue, C. Effect of axial magnetic field in the laser beam welding of stainless steel to aluminum alloy. *Mater. Des.* **2016**, *109*, 146–152. [CrossRef]
31. Torkamany, M.J.; Tahamtan, S.; Sabbaghzadeh, J. Dissimilar welding of carbon steel to 5754 aluminum alloy by Nd:YAG pulsed laser. *Mater. Des.* **2010**, *31*, 458–465. [CrossRef]
32. Chen, S.; Huang, J.; Ma, K.; Zhang, H.; Zhao, X. Influence of a Ni-foil interlayer on Fe/Al dissimilar joint by laser penetration welding. *Mater. Lett.* **2012**, *79*, 296–299. [CrossRef]
33. Cao, X.; Zhou, X.; Wang, H.; Luo, Z.; Duan, J. Microstructures and mechanical properties of laser offset welded 5052 aluminum to press-hardened steel. *J. Mater. Res. Technol.* **2020**, *9*, 5378–5390. [CrossRef]
34. Wei, P.; Chen, Z.; Wang, D.; Zhang, R.; Li, X.; Zhang, F.; Sun, K.; Lei, Y. Effect of laser cleaning on mechanical properties of laser lap welded joint of SUS310S stainless steel and 6061 aluminum alloy. *Mater. Lett.* **2021**, *291*, 129549. [CrossRef]
35. Asirvatham, M.C.; Collins, S.; Masters, I. Laser wobble welding of steel to Aluminium busbar joints for Li-ion battery packs. *Opt. Laser Technol.* **2022**, *151*, 108000. [CrossRef]
36. Sierra, G.; Peyre, P.; Beaume, F.D.; Stuart, D.; Fras, G. Steel to aluminium braze welding by laser process with Al-12Si filler wire. *Sci. Technol. Weld. Join.* **2008**, *13*, 430–437. [CrossRef]
37. Zhang, M.J.; Chen, G.Y.; Zhang, Y.; Wu, K.R. Research on microstructure and mechanical properties of laser keyhole welding-brazing of automotive galvanized steel to aluminum alloy. *Mater. Des.* **2013**, *45*, 24–30. [CrossRef]
38. Xia, H.; Zhao, X.; Tan, C.; Chen, B.; Song, X.; Li, L. Effect of Si content on the interfacial reactions in laser welded-brazed Al/steel dissimilar butted joint. *J. Mater. Process. Technol.* **2018**, *258*, 9–21. [CrossRef]
39. Xia, C.Z.; Li, Y.J.; Wang, J.; Ma, H.J. Microstructure and phase constitution near interface of Cu/Al vacuum brazing. *Mater. Sci. Technol.* **2007**, *23*, 815–818. [CrossRef]
40. Lee, S.J.; Nakamura, H.; Kawahito, Y.; Katayama, S. Effect of welding speed on microstructural and mechanical properties of laser lap weld joints in dissimilar Al and Cu sheets. *Sci. Technol. Weld. Join.* **2014**, *19*, 111–118. [CrossRef]
41. Stritt, P.; Hagenlocher, C.; Kizler, C.; Weber, R.; Rüttimann, C.; Graf, T. Laser spot welding of copper-aluminum joints using a pulsed dual wavelength laser at 532 and 1064 nm. *Phys. Procedia* **2014**, *56*, 759–767. [CrossRef]
42. Mys, I.; Schmidt, M. Laser micro welding of copper and aluminum. In *Laser-Based Micropackaging;* SPIE: Bellingham WA, USA, 2006; p. 610703. [CrossRef]
43. Messler, R.W. *Joining of Materials and Structures;* Chapter 15—Joining Dissimilar Material Combinations; Elsevier: Amsterdam, The Netherlands, 2004; pp. 697–741. [CrossRef]
44. Li, Y.; Zou, W.; Lee, B.; Babkin, A.; Chang, Y. Research progress of aluminum alloy welding technology. *Int. J. Adv. Manuf. Technol.* **2020**, *109*, 1207–1218. [CrossRef]
45. Heideman, R.; Johnson, C.; Kou, S. Metallurgical analysis of Al/Cu friction stir spot welding. *Sci. Technol. Weld. Join.* **2010**, *15*, 597–604. [CrossRef]
46. Guo, Y.; Qiao, G.; Jian, W.; Zhi, X. Microstructure and tensile behavior of Cu-Al multi-layered composites prepared by plasma activated sintering. *Mater. Sci. Eng. A* **2010**, *527*, 5234–5240. [CrossRef]
47. Abbasi, M.; Taheri, A.K.; Salehi, M.T. Growth Rate of Intermetallic Compounds in Al/Cu Bimetal Produced by Cold Roll Welding Process. 2001. Available online: www.elsevier.com/locate/jallcom (accessed on 19 June 2023).
48. Mai, T.A.; Spowage, A.C. Characterisation of dissimilar joints in laser welding of steel-kovar, copper-steel and copper-aluminium. *Mater. Sci. Eng. A* **2004**, *374*, 224–233. [CrossRef]
49. Ali, S.; Shin, J. In-Depth Characterization of Laser-Welded Aluminum-and-Copper Dissimilar Joint for Electric Vehicle Battery Connections. *Materials* **2022**, *15*, 7463. [CrossRef]
50. Lee, S.J.; Choi, K.D.; Lee, S.J.; Shin, D.S.; Jung, J.P. Welding Properties of Dissimilar Al-Cu Thin Plate by a Single-Mode Fiber Laser. *Metals* **2022**, *12*, 1957. [CrossRef]
51. Hailat, M.M.; Mian, A.; Chaudhury, Z.A.; Newaz, G.; Patwa, R.; Herfurth, H.J. Laser micro-welding of aluminum and copper with and without tin foil alloy. *Microsyst. Technol.* **2012**, *18*, 103–112. [CrossRef]
52. Xue, Z.; Hu, S.; Shen, J.; Zuo, D.; Kannatey-Asibu, E. Microstructure characterization and mechanical properties of laser-welded copper and aluminum lap joint. *J. Laser Appl.* **2014**, *26*, 012002. [CrossRef]
53. Weigl, M.; Albert, F.; Schmidt, M. Enhancing the ductility of laser-welde copper-aluminum connections by using adapted filler materia. *Phys. Procedia* **2011**, *12*, 332–338. [CrossRef]
54. Sun, Z.; Ion, J.C. Review Laser welding of dissimilar metal combinations. *J. Mater. Sci.* **1995**, *30*, 4205–4214. [CrossRef]
55. Shi, R.P.; Wang, C.P.; Wheeler, D.; Liu, X.J.; Wang, Y. Formation mechanisms of self-organized core/shell and core/shell/corona microstructures in liquid droplets of immiscible alloys. *Acta Mater.* **2013**, *61*, 1229–1243. [CrossRef]
56. Phanikumar, G.; Manjini, S.; Dutta, P.; Mazumder, J.; Chattopadhyay, K. Characterization of a Continuous CO_2 Laser-Welded Fe-Cu Dissimilar Couple. *Met. Mater. Trans. A* **2005**, *36*, 2137–2147. [CrossRef]
57. Velu, M.; Bhat, S. Metallurgical and mechanical examinations of steel-copper joints arc welded using bronze and nickel-base superalloy filler materials. *Mater. Des.* **2013**, *47*, 793–809. [CrossRef]
58. Joshi, G.R.; Badheka, V.J. Processing of bimetallic steel-copper joint by laser beam welding. *Mater. Manuf. Process.* **2019**, *34*, 1232–1242. [CrossRef]

59. Meng, Y.; Li, X.; Gao, M.; Zeng, X. Microstructures and mechanical properties of laser-arc hybrid welded dissimilar pure copper to stainless steel. *Opt. Laser Technol.* **2019**, *111*, 140–145. [CrossRef]
60. Chen, S.; Huang, J.; Xia, J.; Zhang, H.; Zhao, X. Microstructural characteristics of a stainless steel/copper dissimilar joint made by laser welding. *Met. Mater. Trans. A* **2013**, *44*, 3690–3696. [CrossRef]
61. Yan, F.; Qin, Y.; Tang, B.; Zhou, Y.; Gao, Z.; Hu, Y.; Hu, C.; Xiao, Z.; Xiao, Z.; Wang, C. Effects of beam oscillation on microstructural characteristics and mechanical properties in laser welded steel-copper joints. *Opt. Laser Technol.* **2022**, *148*, 107739. [CrossRef]
62. Mannucci, A.; Tomashchuk, I.; Vignal, V.; Sallamand, P.; Duband, M. Parametric study of laser welding of copper to austenitic stainless steel. *Procedia CIRP* **2018**, *74*, 450–455. [CrossRef]
63. Li, J.; Cai, Y.; Yan, F.; Wang, C.; Zhu, Z.; Hu, C. Porosity and liquation cracking of dissimilar Nd:YAG laser welding of SUS304 stainless steel to T2 copper. *Opt. Laser Technol.* **2020**, *122*, 105881. [CrossRef]
64. Shen, H.; Gupta, M.C. Nd:yttritium–aluminum–garnet laser welding of copper to stainless steel. *J. Laser Appl.* **2004**, *16*, 2–8. [CrossRef]
65. Kirchhoff, M. Laser applications in battery production—From cutting foils to welding the case. In Proceedings of the 2013 3rd International Electric Drives Production Conference (EDPC), Nuremberg, Germany, 29–30 October 2013. [CrossRef]
66. Kawahito, Y.; Matsumoto, N.; Abe, Y.; Katayama, S. Relationship of laser absorption to keyhole behavior in high power fiber laser welding of stainless steel and aluminum alloy. *J. Mater. Process. Technol.* **2011**, *211*, 1563–1568. [CrossRef]
67. Katayama, S.; Nagayama, H.; Mizutani, M.; Kawahito, Y. Fibre laser welding of aluminium alloy. *Weld. Int.* **2009**, *23*, 744–752. [CrossRef]
68. Xiao, R. Laser beam welding of aluminum alloys. In *Lasers in Material Processing and Manufacturing III*; SPIE: Bellingham, WA, USA, 2007; p. 682505. [CrossRef]
69. Xiao, R.; Zhang, X. Problems and issues in laser beam welding of aluminum-lithium alloys. *J. Manuf. Process.* **2014**, *16*, 166–175. [CrossRef]
70. Wu, Q.; Xiao, R.S.; Zou, J.L.; Xu, J.J. Weld formation mechanism during fiber laser welding of aluminum alloys with focus rotation and vertical oscillation. *J. Manuf. Process.* **2018**, *36*, 149–154. [CrossRef]
71. Möller, M.; Vogt, S. Novel multi focus optic approach for gas-tight welding of aluminum alloys in e-mobility. *Procedia CIRP* **2022**, *111*, 774–777. [CrossRef]
72. Prieto, C.; Vaamonde, E.; Diego-Vallejo, D.; Jimenez, J.; Urbach, B.; Vidne, Y.; Shekel, E. Dynamic laser beam shaping for laser aluminium welding in e-mobility applications. *Procedia CIRP* **2020**, *94*, 596–600. [CrossRef]
73. Abeyrathna, B.; Zhang, P.; Pereira, M.P.; Wilkosz, D.; Weiss, M. Micro-roll forming of stainless steel bipolar plates for fuel cells. *Int. J. Hydrogen Energy* **2019**, *44*, 3861–3875. [CrossRef]
74. Salgado, M.D.F.; Rodrigues, S.C.; Santos, D.M.; Brandim, A.S.; Lins, V.F. Cyclic oxidation resistance of ferritic stainless steels used in mufflers of automobiles. *Eng. Fail. Anal.* **2017**, *79*, 89–97. [CrossRef]
75. Samaniego-Gámez, O.; Almeraya-Calderón, F.; Chacón-Nava, J.; Maldonado-Bandala, E.; Nieves-Mendoza, D.; Rios, J.P.F.-D.L.; Jáquez-Muñoz, J.M.; Delgado, A.D.; Gaona-Tiburcio, C. Corrosion Behavior of Passivated CUSTOM450 and AM350 Stainless Steels for Aeronautical Applications. *Metals* **2022**, *12*, 666. [CrossRef]
76. Schröder, C.; Wendler, M.; Kreschel, T.; Volkova, O.; Weiß, A. Development of a stainless austenitic nitrogen-alloyed crmnnimo spring steel. *Crystals* **2019**, *9*, 456. [CrossRef]
77. De Baglion, L.; Mendez, J. Low cycle fatigue behavior of a type 304L austenitic stainless steel in air or in vacuum, at 20 °C or at 300 °C: Relative effect of strain rate and environment. *Procedia Eng.* **2010**, *2*, 2171–2179. [CrossRef]
78. Zhang, H.W.; Xu, Y.L.; Chen, L.J.; Wang, X.W.; Wu, Z.Y.; Li, S.; Li, J.; Xiao, X.S. Precipitation Behaviors of a New Antibacterial Maraging Stainless Steel for Medical Instruments. *Met. Mater. Trans. A* **2018**, *49*, 3753–3761. [CrossRef]
79. Alcock, J.A.; Baufeld, B. Diode laser welding of stainless steel 304L. *J. Mater. Process. Technol.* **2017**, *240*, 138–144. [CrossRef]
80. Shankar, V.; Gill, T.P.S.; Mannan, S.L.; Sundaresan, S. Solidification cracking in austenitic stainless steel welds. *Sadhana* **2003**, *28*, 359–382. [CrossRef]
81. Reitz, W. A Review of: 'Welding Metallurgy and Weldability of Stainless Steel'. *Mater. Manuf. Process.* **2006**, *21*, 219. [CrossRef]
82. Nishimoto, K.; Saida, K.; Kiuchi, K.; Nakayama, J. Influence of minor and impurity elements on hot cracking susceptibility of extra high-purity type 310 stainless steels. In *Hot Cracking Phenomena in Welds III*; Springer: Berlin/Heidelberg, Germany, 2011; pp. 183–208. [CrossRef]
83. Almomani, A.; Mourad, A.H.I.; Barsoum, I. Effect of sulfur, phosphorus, silicon, and delta ferrite on weld solidification cracking of AISI 310S austenitic stainless steel. *Eng. Fail. Anal.* **2022**, *139*, 106488. [CrossRef]
84. Lienert, T.J.; Lippold, J.C. Improved weldability diagram for pulsed laser welded austenitic stainless steels. *Sci. Technol. Weld. Join.* **2003**, *8*, 1–9. [CrossRef]
85. Ning, J.; Wen, J.H.; Zhang, L.J.; Na, S.J. Assessment of the universality of duplex stainless steel powder in laser additive repair based on Schaeffler diagram. *Addit. Manuf.* **2022**, *55*, 102864. [CrossRef]
86. Varol, I.; Baeslack, W.A., III; Lippold, J.C. Characterization of Weld Solidification Cracking in a Duplex Stainless Steel. *Metallography* **1989**, *23*, 1–19. [CrossRef]
87. Yan, J.; Gao, M.; Zeng, X. Study on microstructure and mechanical properties of 304 stainless steel joints by TIG, laser and laser-TIG hybrid welding. *Opt. Lasers Eng.* **2010**, *48*, 512–517. [CrossRef]

88. P'ng, D.; Molian, P. Q-switch Nd:YAG laser welding of AISI 304 stainless steel foils. *Mater. Sci. Eng. A* **2008**, *486*, 680–685. [CrossRef]
89. Zhang, M.; Chen, G.; Zhou, Y.; Liao, S. Optimization of deep penetration laser welding of thick stainless steel with a 10 kW fiber laser. *Mater. Des.* **2014**, *53*, 568–576. [CrossRef]

Disclaimer/Publisher's Note: The statements, opinions and data contained in all publications are solely those of the individual author(s) and contributor(s) and not of MDPI and/or the editor(s). MDPI and/or the editor(s) disclaim responsibility for any injury to people or property resulting from any ideas, methods, instructions or products referred to in the content.

coatings

MDPI

Review

Crack Formation Mechanisms and Control Methods of Laser Cladding Coatings: A Review

Mingke Li, Kepeng Huang and Xuemei Yi *

College of Mechanical and Electronic Engineering, Northwest Agriculture and Forestry University, Xianyang 712100, China; limingke@nwafu.edu.cn (M.L.); kepeng_huang@nwafu.edu.cn (K.H.)
* Correspondence: xuemei_yi@nwsuaf.edu.cn; Tel.: +86-29-8709-2391

Abstract: Laser cladding, a novel surface treatment technology, utilizes a high-energy laser beam to melt diverse alloy compositions and form a specialized alloy-cladding layer on the surface of the substrate to enhance its property. However, it can generate substantial residual stresses during the rapid cooling and heating stages, due to inadequate selection of cladding process parameters and disparities in thermophysical properties between the clad layer and substrate material, leading to the formation of various types of cracks. These cracks can significantly impact the quality and performance of the coating. This paper presents a comprehensive review of crack types and their causes in laser cladding coatings, and identifies that three primary sources of residual stresses, thermal stress, organizational stress, and restraint stress, are the fundamental causes of crack formation. The study proposes several strategies to control coating cracks, including optimizing the coating layer material, refining the coating process parameters, incorporating heat treatment, applying auxiliary fields, and utilizing numerical simulations to predict crack initiation and propagation. Additionally, the paper summarizes crack control methods for emerging structural materials and novel preparation processes. Lastly, the paper analyzes the prospects, technical approaches, and key research directions for effectively controlling cracks in laser cladding coatings.

Keywords: laser cladding coating; cracks; residual stress; causes; control methods

1. Introduction

Laser cladding (LC) is an advanced surface modification technology that utilizes a high-energy laser beam to melt the clad material, forming a strong metallurgical bond with the substrate material [1]. This technology offers numerous advantages, including high bond strength, minimal heat-affected zone, low thermal deformation, and low dilution rate [2–4]. It has found extensive applications in aerospace, automotive, and chemical industries, among others [5–7]. The LC system usually consists of a KUKA robot or robot arm, LC head, powder feeder, protective gas, water cooling system, and workbench [8] (Figure 1).

Despite the excellent performance of laser melting coatings, the rapid melting and solidification process can give rise to defects such as cracks, porosity, and inclusions within the coating. Among these defects, cracks are the most common internal issue in the laser cladding process [9,10]. Microscopic cracks within the coating can propagate into macroscopic cracks under working loads, significantly compromising the coating's quality [11]. Consequently, current research focuses on suppressing or eliminating internal cracks in laser cladding coatings, which presents a challenging and active area of study [12,13].

In recent years, scholars have conducted extensive experimental research on the causes and control methods of cracks in laser cladding coatings. Zhang et al. [14] and Galy et al. [15] reviewed the solidification theory of the selective laser melting (SLM) process and the formation mechanisms of the hole and crack defects. Quazi et al. [16] discussed the influence of rare earth additives on the crack sensitivity of coatings. Similarly, Wang et al. [17] and Hu et al. [18] suggested improvement strategies, including optimizing process parameters and employing preheating treatments, to mitigate coating cracks. The formation of cracks in laser cladding coatings is primarily attributed to residual stresses. While existing equipment can measure residual stress, the process is intricate, challenging, and costly. To overcome this limitation, scholars have conducted relevant studies using numerical simulations. Fang et al. [19] reviewed physical models of residual stresses for defects in selective laser melting (SLM) and machined parts, analyzed the advantages and disadvantages of mainstream models, and proposed conceptual methods to enhance residual stress management. Sanaei et al. [20] discussed characterization methods and statistical analysis of coating defects in additive manufacturing (AM), summarizing the effects of process parameters and post-processing on defects. Cheng et al. [21] reviewed the simulation techniques for grain growth mechanism, temperature, and stress distribution in the melt pool directly related to defect formation in the laser metal deposition (LMD) technique. Additionally, the defect suppression methods and the performance improvement methods of filled layers in LMD technology are presented.

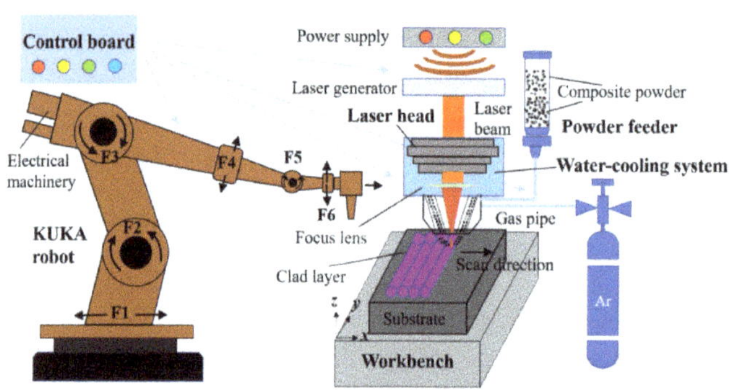

Figure 1. Schematic diagram of the laser cladding system.

In summary, while numerous scholars have provided an overview of crack generation mechanisms and control methods for coatings, there has been limited analysis of the types, formation mechanisms, and control methods of various internal cracks in laser melting coatings. This article comprehensively lists the types of internal cracks in laser cladding coatings from the perspectives of experiments and numerical simulations. The formation mechanisms of coating cracks are elucidated, and a comparative analysis of different control methods for coating cracks is provided. Furthermore, future control methods and new technologies for addressing cracks in laser cladding coatings are discussed, offering valuable insights for crack control and improvement in coating quality in laser cladding processes.

2. Types and Causes of Cracks

2.1. Types of Cracks

Laser-clad coatings exhibit various types of cracks, which can be attributed to the selection of different cladding materials and process parameters. To effectively address the cracking issue in laser-clad coatings, it is crucial to identify the types of coating cracks and understand their causes. To address the issue of the types of cracks in laser-clad coatings, they can be classified as hot and cold cracks according to the time, temperature, and fracture characteristics of the coating cracks, and most of the coating cracks are cold cracks [22].

Hot cracks mainly occur above the solidification temperature line, and their sections have a distinct oxidation color and no metallic luster (Figure 2a). For example, most of the common types of cracks in melt layers, such as austenitic stainless steels, are hot cracks [23]. Hot cracks sprout from hot tearing and are easily influenced by microstructure [24]. Partition of low melting point elements such as Si and C in the molten layer [25,26], inhomogeneous distribution of coarse and brittle phases of compounds and impurities [27], and disordered grain growth are the main factors for the formation of hot cracks [28]. This is due to the fact that the liquid metal in the melt pool has a very high temperature, so the liquid metal is componentlessly subcooled at the beginning of solidification, and therefore large tensile stresses are formed at the coarse eutectic tissue at low melting points. The tensile stress pulls the solidified tissue partially along the grain boundary, resulting in not enough liquid phase to fill the tissue gap, thus producing cracks. Therefore, thermal-type cracks mostly show cracking characteristics along the grain [29].

Cold cracks mainly occur below the solidification temperature line, and the fractured section of cold cracks appears relatively smooth with a metallic luster (Figure 2b). Cold cracks often involve secondary crack generation, indicating a brittle fracture behavior. For instance, nickel-based alloy powder coatings commonly exhibit cold cracks [30]. Improper selection of melting process parameters and excessive thermal gradients in the melt pool are the main factors contributing to cold crack formation. This is due to the susceptibility of martensitic phase transformation at the solidification temperature line [31], leading to the hardening of the coating. The differences in thermal–physical parameters between the cladding material and the substrate material, coupled with the cooling and solidification process, result in the generation of large residual thermal stresses. When these residual thermal stresses surpass the tensile strength limit of the material, cold cracks occur [31–33]. Consequently, cold cracks typically exhibit crack propagation through the crystal structure [34].

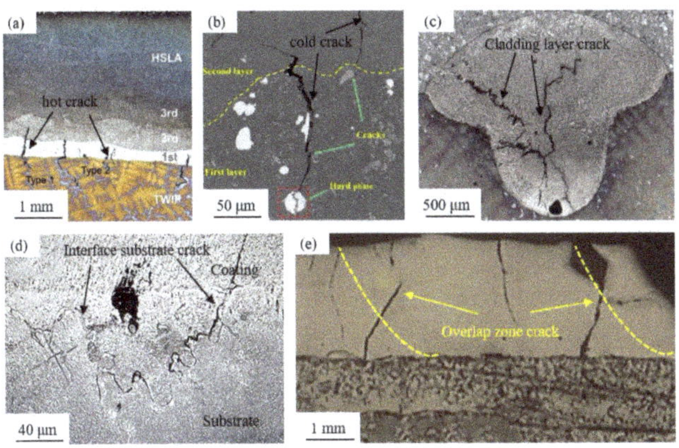

Figure 2. Different types of crack morphology. (**a**) The hot cracks in HSLA powder samples (the laser power is 2900 W and the scan speed is 10 mm/s) [25]; (**b**) the cold cracks in Ni-Cu alloy sample (the laser power is 5000 W and the scan speed is 30 mm/s) [35]; (**c**) cladding layer crack in ZhS32 alloy (the laser power is 600 W and the scan speed is 9 mm/s) [36]; (**d**) interface substrate crack in nickel-based K477A (the laser power is 576 W and the scan speed is 4 mm/s) [37]; (**e**) overlap zone crack in Ni60 (the laser power is 3200 W and the scan speed is 416.7 mm/s) [38].

Depending on their location, coating cracks can be classified as fusion layer cracks, interface substrate cracks, and overlap zone cracks [37–39].

Cladding layer crack is caused by rapid cooling during the solidification process, leading to significant thermal stress. Uneven stress distribution resulting from the uneven mixing of cladding material or the presence of impurities can also contribute to the cracking of the cladding layer (Figure 2c). When cracks form within the clad layer, they often initiate near hard-phase particles and propagate vertically through the coating. In some cases, the cracks even extend into the substrate. When cracks reach the surface, they exhibit a "herringbone" or lattice-like pattern [34,39].

Interface substrate crack primarily arises from the excessive difference in thermal expansion coefficient and Young's modulus between the substrate and the molten cladding layer. This difference creates thermal stresses under large temperature gradients during solidification and cooling. Additionally, the molten metal liquid experiences restraint stress from the substrate during thermal expansion and cooling contraction. As a result of the combined effects of thermal and restraint stresses, cracking occurs at the interface between the clad layer and the substrate, gradually propagating to the surface of the molten cladding layer (Figure 2d) [37], forming interface substrate cracks.

Overlap zone cracks predominantly occur in the lap zone as a result of an unreasonable selection of the overlap rate. This leads to the accumulation of heat, an increase in temperature gradient, and elevated thermal stress. Consequently, the grain growth time is prolonged, resulting in coarser grains, as well as the formation of pores or impurities. The expansion of these defects eventually leads to the development of lap zone cracks (Figure 2e). Moreover, the low yield strength and tensile rate of the clad material make it more susceptible to cracking and expansion under combined stress. Consequently, overlap zone cracks tend to propagate throughout the entire fusion cladding layer once they form [40,41].

In summary, hot cracks are mainly caused by hot tearing and are significantly influenced by the microstructure of the coating. The presence of coarse brittle phases and impurities in the molten clad layer, along with thermal cycling and stress concentration due to the shaped orientation of grain boundaries, are the key factors contributing to the formation of hot cracks. Cold cracks, on the other hand, are mainly a consequence of improper selection of process parameters and excessive temperature gradients, resulting in brittle fractures due to tensile stresses exceeding the tensile strength of the molten material. Cracks in the cladding layer arise from the uneven distribution of coating tissue and a substantial difference in the thermal expansion coefficient of the cladding layer. Interface substrate cracks primarily occur due to the excessive temperature gradient between the substrate and the clad layer. Overlap zone cracks, finally, are predominantly caused by inappropriate overlap rate selection. These different types of coating cracks can be analyzed based on their causes of formation, with the strain generated during the melting process being greater than the plastic strain of the molten layer itself, serving as the fundamental cause of coating cracks.

2.2. Causes of Cracks

The laser cladding process is very complex, and the coating generates significant stresses during the cladding and cooling process. When the stresses in the coating exceed the yield limit of the coating material, they can lead to cracking of the coating. The formation of cracks in laser-clad coatings can be attributed to the presence of residual stresses, which can be categorized into three primary types: thermal stresses, organizational stresses, and restraint stresses (Figure 3) [42–44].

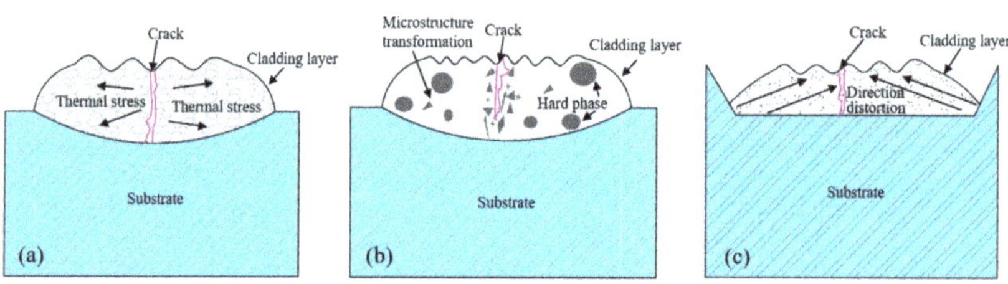

Figure 3. Schematic of principle of three different stress: (**a**) thermal stress; (**b**) organizational stress; and (**c**) restraint stress.

2.2.1. Thermal Stress

The presence of thermal stress in laser-clad coatings can be attributed to the disparate modulus of elasticity and coefficient of thermal expansion between the substrate and the clad material. This results in varying rates of thermal expansion and cooling contraction within the clad layer. When a temperature gradient exists, the clad layer experiences stress, known as thermal stress (Figure 3a) [45]. If the thermal stress surpasses the material's yield limit, it can give rise to cracks in the coating. The calculation of thermal stress involves determining the extent of the stress [45].

$$\sigma_T = \frac{E \times \Delta\alpha \times \Delta T}{1 - \nu} \tag{1}$$

E is the elastic modulus of the cladding material, $\Delta\alpha$ is the difference in thermal expansion coefficient between the cladding layer and substrate material, ΔT is the difference between cladding temperature and room temperature, and v is Poisson's ratio of the cladding layer.

The relationship expressed in Equation (1) reveals that the Poisson's ratio of the molten layer decreases as the thermal expansion coefficient difference between the molten layer and the substrate material increases. Consequently, a larger temperature difference leads to greater thermal stress, making the coating more susceptible to cracking [46]. Because the substrate is difficult to replace in the selection process, and the coefficient of thermal expansion and Young's modulus of the cladding material also differ greatly, the selection of cladding material is particularly important. For instance, when the coating material consists of ceramic particles and the substrate is made of a metal alloy, the distinct thermophysical properties of these materials can cause cracking if the number of ceramic particles added is not carefully chosen. To address this issue, the composition of the cladding layer is commonly adjusted to alleviate the impact of thermal stress [47].

2.2.2. Organizational Stress

During the solidification and crystallization process of the liquid metal within the molten pool, a rearrangement of the physical phase structure takes place, resulting in the generation of internal stress known as organizational stress (Figure 3b) [48]. As thermal cycling progresses, the organizational stress accumulates until it reaches the yield strength of the material. This accumulation of stress can lead to tissue damage, the formation of microscopic defects, and even the development of microcracks, ultimately resulting in brittle fractures. Griffith introduced and refined a model that relates brittle fracture strength to material properties and damage [48], which can be expressed as follows:

$$\sigma_f = \sqrt{\frac{E(2\gamma + \gamma_p)}{\pi C_z}} \tag{2}$$

E is the modulus of elasticity, γ is the material surface energy, γp is the microcracks or micro defects expanding the plastic work per unit length, C_z is the size of the microcracks and micro defects.

The fracture surface energy γ is about $\frac{aE}{100}$, a is the lattice distance of the crystal; the plastic work γp is 2~3 orders of magnitude larger than γ.

The fracture strength of a material can be estimated by considering the size of the microcrack and the average lattice distance, although the specific value varies depending on the crystallographic system. For instance, in laser-fused coatings containing hard phases such as Cr_7C_3 and $Cr_{23}C_6$, when the size of the microdamage crack within the coating ranges from 3 to 5 μm, the calculated fracture strength (denoted as "σ_f") is approximately 4.4 GPa [49]. When the stress in the coating exceeds this fracture strength, the microcrack extends and develops into macroscopic cracks. Hard phases are generally added to the coating to improve its performance, resulting in poor fluidity of the coating and uneven powder mixing. Therefore, the coating, after melting, is prone to elemental segregation. The presence of elemental segregation contributes to an inhomogeneous phase transition structure, resulting in stress concentration and elevated organizational stresses. Ramakrishnan et al. [50] proposed that Inconel 738 coating cracking occurs due to the micro-segregation of aluminum (Al) and titanium (Ti) elements, as well as the presence of low melting point crack boundaries.

The concept of micro-bias is directly associated with the diffusion coefficient of liquid alloying elements in the melt pool. This diffusion coefficient can be determined using Formula (3) [51].

$$K = A \times e^{\frac{E}{RT}} \tag{3}$$

K is the rate coefficient, E is the activation energy, T is the absolute temperature, A is the frequency factor, and R is the ideal gas law constant.

The relationship expressed in Formula (3) reveals that an increase in the cooling rate leads to a decrease in the diffusion coefficient of elements, resulting in reduced microscopic segregation and decreased crack sensitivity. To mitigate these effects and minimize organizational stress caused by uneven phase transformation, various approaches can be employed. These include implementing auxiliary fields, refining the size of the hard phase, or selecting appropriate process parameters. These measures help slow down elemental segregation and promote a more uniform phase distribution within the material.

2.2.3. Constraint Stress

During laser cladding, as the molten cladding layer undergoes heating and expansion or cooling and contraction, hindering stress arises from the un-melted portion on the melted part, referred to as restraint stress (Figure 3c) [52]. Two types of restraint stresses can be identified: the first is compressive stress resulting from the thermal expansion of the initially melted material in the melt pool, constrained by the colder surrounding substrate; the second is tensile stress generated as the molten liquid metal is held by other colder parts of the substrate during condensation and cooling shrinkage.

The constraint stress mainly comes from the constraints of the matrix, so the constraint stress at the edge of the cladding layer is relatively small and less prone to cracking. Research indicates that the middle section of the molten cladding layer is particularly susceptible to cracking due to heat accumulation and limited heat dissipation area, thus requiring additional constraints [53]. Appropriate heat treatment techniques can help reduce restraint stress [54]. In addition, the use of low transformation temperature (LTT) alloys can effectively reduce the accumulation of tensile stress. This is because the low transformation temperature alloy can use the expansion of martensite transformation to offset part or all of the heat shrinkage, thereby reducing the residual tensile stress [55]. However, it is important to note that the melting and cooling behavior of the laser cladding pool is highly complex, and residual stresses are challenging to eliminate entirely. The non-uniform distribution of temperature, stress, and flow fields within the melt pool exacerbates residual stresses, especially when there is a significant temperature gradient or a mismatch in the thermal expansion coefficient and elastic modulus between the molten material and the substrate. Consequently, regulating thermal stress, organizational stress, and restraint stress during the laser cladding coating process, minimizing the individual or interactive effects of these stresses, and preventing or eliminating coating cracking remain significant challenges in current research.

3. Crack Control Method

To address the issue of cracking in laser-clad coatings, this paper provides a comprehensive review of several strategies aimed at controlling crack formation. These methods include the preferential selection of clad layer materials [56–58], optimization of coating process parameters [59], implementation of increased heat treatment [46], utilization of auxiliary fields [60,61], and the use of numerical simulations [62]. Furthermore, this review highlights recent advancements and novel approaches in the field, summarizing the progress made in controlling coating cracking in recent years.

3.1. Preferred Cladding Material

3.1.1. Control of Hard-Phase Content in the Cladding Layer

There are many applications of laser cladding, among which the most prominent application is the preparation of surface-strengthening coatings to improve the hardness and wear resistance of parts. It has been demonstrated that the proper addition of hard-phase particles with high melting points, such as WC, TiC, and ZrO_2, can effectively refine the grains of the coating. This refinement reduces organizational stresses caused by elemental segregation and lowers the susceptibility of the coating to cracking [63]. For instance, Zhao et al. [64] observed that Ni-based WC-CeO_2 coatings exhibited smaller secondary dendrite spacing and were free from cracks compared to Ni-based CeO_2 coatings. This improvement can be attributed to the addition of WC hard-phase particles, which increase the number of carbides in the eutectic structure and refines the grains, thereby enhancing the coating's toughness and reducing its crack susceptibility. However, it is crucial to control the number of hard-phase particles added, as excessive differences in thermal expansion coefficient and elastic modulus between the coating and substrate can lead to coating cracking. Shen et al. [65] investigated the cracking of NiCrSiBC-WC composite coatings with WC mass ratios of 40%, 50%, and 60% on carbon steel substrates (Figure 4a). The study revealed that the composite coatings with 40 wt.% WC exhibited no cracks, while those with 50 and 60 wt.% WC showed cracks, with significant penetration cracks observed in the middle due to the presence of large grains of WC.

Similar observations have been made regarding the effect of hard-phase particle content on the surface quality and structure of coatings. Shen et al. [66] studied the cracking of NiCrSiBC-WC composite coatings and found that coatings with 30–40 wt.% WC exhibited no cracking or fragmentation of WC particles, whereas coatings with 50–60 wt.% WC exhibited significant deterioration and increased cracking. This can be attributed to the fact that the secondary precipitation phases of WC, such as bulk carbides and rod carbides (laminated carbides), have poor wetting properties with nickel and are highly susceptible to stress concentration, leading to cracking [46]. Similarly, Liu et al. [67] investigated the effect of TiC volume fraction on the microstructure and tensile properties of TiC/TA15 titanium-based composite coatings. They discovered that composites with 5 vol.% TiC exhibited improved strength, whereas the tensile properties deteriorated as the TiC volume fraction increased, resulting in coating cracking. Similarly, Ignat et al. [47] found that when the volume fraction of ZrO_2 in the coating exceeded 15%, the surface quality of the coating deteriorated with the appearance of cracks.

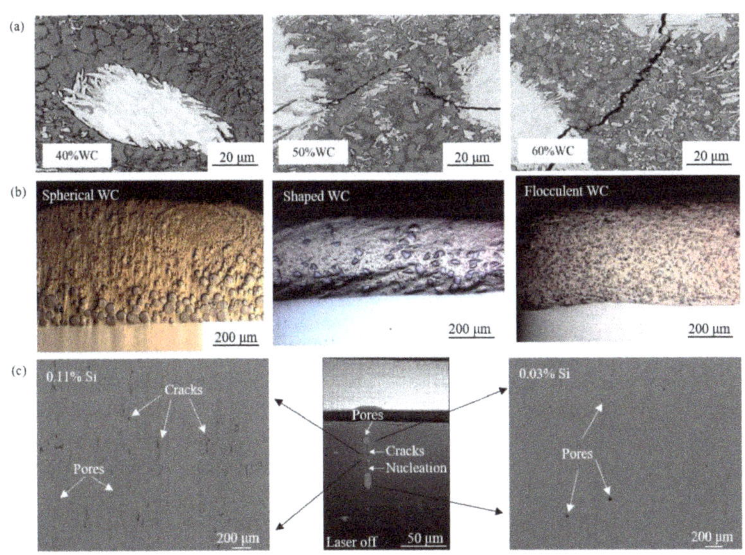

Figure 4. The effect of cladding materials on the microscopic morphology of coatings. (**a**) Micromorphology of NiCrBSi-WC coatings with different WC contents [65]; (**b**) microscopic photos of different types of WC coatings in Ni60-30 wt.% WC composite coatings [68]; and (**c**) microscopic photos of coatings with different silicon contents in IN738LC coatings [69].

In summary, when incorporating ceramic hard-phase particles into the coating, it is important to limit the hard-phase content. The commonly used WC ceramic particle content should be kept below 50 wt.%, and the content of other less dense hard-phase particles should also be controlled. Otherwise, excessive stress may lead to cracking in the coating.

3.1.2. Preferred Hard-Phase Particle Type

Hard-phase particles can be categorized into microns and nanoparticles based on their particle size [70]. They can also be classified as spherical, shaped, or flocculent based on their shape [68]. Furthermore, hard-phase particles can be classified according to their structure as mechanically mixed or encapsulated [71]. The addition of different types of hard-phase particles results in variations in coating properties and the occurrence of coating cracks, and the choice of preferred hard-phase particles can enhance the crack resistance of the coating.

Regarding the impact of hard-phase particle size on coating cracks, researchers have found that larger particles can impede crack expansion. However, when the particle size is excessively large, it can lead to stress concentration and subsequent coating cracking [72]. He et al. [73] investigated the influence of micron and nano-sized TiC particles on the structure and properties of TiC-TiAl composite coatings. They discovered that the coating containing 20 wt.% micron-sized TiC had 14 cracks, while the coating with 10 wt.% nano-sized TiC had only 6 cracks, and the coating with 20 wt.% nano-sized TiC had a mere 3 cracks. Wang et al. [74] observed the elimination of cracks near the substrate interface upon the addition of 1 wt.% nano-sized Al_2O_3 to NiCoCrAlY high-temperature protective coatings. Li et al. [75] further demonstrated that nanoceramic particles not only eliminated coating cracks but also transformed columnar crystals into equiaxed crystals with a more uniform and dense structure.

In terms of the influence of hard-phase particle shape on coating cracking, Zhang et al. [68] investigated the effect of spherical, shaped, and flocculent WC particles (30 wt.%) on Ni60 coatings. They found that spherical WC particles mainly accumulated in the lower-middle part of the coating, while shaped WC particles showed a similar distribution pattern. In contrast, smaller flocculent WC particles were uniformly dispersed throughout the coating (Figure 4b), and coatings prepared with flocculent WC particles exhibited a smaller tendency to crack.

Regarding the impact of hard-phase particle structure on coating cracks, mechanically mixed WC particle coatings in a Ni-based substrate commonly exhibit cracking, whereas coatings prepared with overlapping particles such as WC-Co and WC-Ni demonstrate overall improved performance. Zhang et al. [76] observed that four Ni-Cu/WC-12Co composite coatings with WC-12Co content ranging from 0 to 30 wt.% showed no cracks. Similarly, the WC-Ni coating prepared by Tehrani et al. [77] was not only crack-free but also 1.5 times tougher than the WC-Co thermal spray coating.

In summary, the proper addition of nano-sized hard-phase particles can refine the grain structure of the coating and enhance its toughness. Flocculent hard-phase particles can be evenly distributed within the coating, thereby preventing stress concentration caused by particle aggregation. Encapsulated hard-phase particles possess good mobility and wettability, improving the coating's toughness and playing a stronger role in reinforcement while impeding crack expansion. Therefore, when selecting the type of hard-phase particles, consideration should be given to their particle size, shape, and structure.

3.1.3. Adding Alloying Elements

Studies have demonstrated that the presence of excessive residual stress in the coating can lead to brittle cracking. However, the addition of alloying elements such as Co, Cr, Si, and V_2O_5 to the coating can enhance fracture toughness and reduce its susceptibility to cracking.

In a study conducted by Dai et al. [78] on the effect of Co content on the cracking susceptibility of WC/Fe laser-clad layers, it was observed that the toughness of the coating increased with the addition of Co. When 8 wt.% Co was added, the toughness reached nearly 1100 MPa, which was 1.3 times higher than the coating without Co. Hence, the addition of the Co element effectively mitigates the tendency of the coating to undergo brittle fracture and exhibits favorable toughening and plasticizing effects. Qi et al. [79] investigated the mechanism behind the improvement in coating cracking through the addition of the Cr element. They found that the presence of Cr gradually refined the coating grains (Figure 4d). Moreover, the addition of the Cr element positively influenced the reduction in the elastic modulus of the clad layer, thereby enhancing the coating's toughness and alleviating thermal stress and solidification shrinkage in the clad layer, ultimately reducing the susceptibility to coating cracking. Sun et al. [69] observed a crack density of 0.85 mm/mm for an IN738LC high-temperature alloy coating with a 0.11 wt.% Si content, whereas the sample with 0.03 wt.% Si content exhibited minimal cracking and only a few pores (Figure 4c), consistent with the findings of Cloots et al. [24]. Wang et al. [23] noted that high volume fraction and uneven distribution of hard phases such as chromium boride and carbide in NiCrBSiC coatings resulted in thermal stress concentration. However, the addition of V_2O_5 led to grain refinement, improved coating toughness, reduced crack sensitivity, and the elimination of cracking when added at 5 wt.%.

In summary, the addition of alloying elements aims to enhance the toughness of the coating, increase its yield limit, and transition it from a brittle to more ductile behavior. Furthermore, it promotes grain refinement and uniform distribution within the coating to avoid stress concentration, thereby reducing the coating's susceptibility to cracking.

3.1.4. Addition of Rare Earth Oxides

Studies have revealed that the incorporation of rare earth elements, such as La, Ce, and Y, which exhibit surface active properties, can effectively reduce the surface tension of molten metal during coating formation. This reduction in surface tension improves the mobility of the melt pool, consequently minimizing defects such as porosity and cracks in the coating [16]. For instance, Sun et al. [80] synthesized (Ti, Nb) C/Ni coatings through in situ synthesis and investigated the impact of CeO_2 additions ranging from 0 % to 9 wt.% on the coatings' microstructure and defect formation. It was observed that coatings without CeO_2 or with excessive CeO_2 additions displayed more pores and cracks, whereas the addition of 5 wt.% CeO_2 significantly improved the microstructure.

Moreover, rare earth elements undergo chemical reactions with impurity elements (e.g., S, O, Si, and N) upon thermal decomposition, leading to the formation of stable high melting point compounds that serve as crystalline nuclei. This process increases the number of nucleation sites, resulting in a greater number of grains and a refined coating structure. Consequently, the toughness of the coating is enhanced, and defects within the coating are reduced [81]. Table 1 presents the impact of different rare earth oxides on the improvement in coating cracks.

Table 1. Improvement effect of different types of rare earth oxides on coating cracks.

Coating/Substrate	Rare Earth Oxide wt.%	Improvement Effect
Fe-based/5CrNiMo [82]	1%/2%/3% Y_2O_3	A 2% Y_2O_3 coating was added without cracks or pores
TiB-TiC/Ti6Al4V [83]	2% Y_2O_3	The average fracture toughness was increased from 8.32 MPa $m^{1/2}$ to 17.36 MPa $m^{1/2}$, and the cracking resistance was improved
Ni-based/45 steel [84]	0.1%–1% La_2O_3	The addition amount was 0.6%; the coating had no cracks, and the microstructure was refined
Ni-60%WC/ASTM A36 [85]	0.5%–2% La_2O_3	A 1% La_2O_3 was added; the coating was tight and defect free When the addition amount was greater than 1.5%, the oxygen content increased, and defects increased
Ti/Ti6Al4V [86]	1% CeO_2 and 1% Y_2O_3	Refined the coating structure and ensured that the coating had no defects
CrTi4-TiCx/Ti6Al4V [87]	1%–4% CeO_2	The addition amount is 2%; the coating has no cracks
Ni60A/TC4 [88]	1%–4% CeO_2	The fluidity of the molten pool improved, and the addition amount is 3%; there are no obvious cracks in the coating
Ni60/6063Al [89]	5% Y_2O_3, CeO_2, and La_2O_3	Dense organization, refined grain size, without obvious cracks and pores

In summary, the selection and quantity of rare earth oxides added vary depending on the specific molten coating material, typically within a controlled range of 5 wt.%. Notably, when the addition of rare earth oxides falls between 0.5 and 2 wt.%, the resulting coating exhibits a dense structure with small grains and minimal observable defects. Excessive addition of rare earth oxides leads to increased oxygen content within the coating and the introduction of impurities at grain boundaries, thereby exacerbating defect formation. Hence, strict control over the addition of rare earth oxides is crucial.

3.1.5. Adding a Transition Layer

Transition layers in laser cladding coatings can be classified into two categories. The first category is the buffer layer, which possesses thermophysical properties like the substrate. This buffer layer acts as an intermediary between the cladding layer and the substrate, reducing the residual stress in the coating and minimizing the dilution effect caused by the substrate [90]. The second category is step coating. In cases where the melt layer contains a high concentration of hard-phase particles, a layer-by-layer incremental approach can be employed to incorporate the hard-phase particles and alleviate the residual stresses between different layers of the coating [91,92].

In studies focusing on buffer layers for stress reduction in coatings, Cao et al. [93] successfully prepared Cr_3C_2-reinforced Ni60A-Ag self-lubricating composite coatings on a copper alloy substrate using a NiCr buffer layer. This approach effectively addressed issues related to segregation and cracks caused by coarse Cr_7C_3 columnar grains in single-layer composite coatings. Similarly, through pre-coating a CoCrFeMnNi melt layer on the substrate, Zhang et al. [94] achieved the successful formation of a 5–20 wt.% TiC composite-reinforced coating, which reduced stress concentration and enhanced the coating's resistance to cracking. Thawari et al. [95] utilized an Inconel 625 buffer layer to deposit a wear-resistant Stellite 6 coating on an SS316 substrate, resulting in a coating with reduced defects such as porosity and cracks. The buffer layer prevented direct contact between Stellite 6 and the substrate, thereby minimizing the temperature gradient (G) and slowing down heat loss in the molten layer.

Regarding the reduction in stress in step fusion coatings, Paul et al. [71] employed a pulsed Nd: YAG laser to deposit multilayer heavy gradient fusion layers on a mild steel substrate. This approach effectively addressed issues such as cracks, porosity, poor bonding, and partial melting of WC particles in the Co substrate. Wang et al. [96] utilized a gradient combination design in the preparation of WC-enhanced nickel-based gradient composite coatings, significantly reducing the susceptibility to cracking in coatings with a high concentration of hard-phase particles. Similarly, Li et al. [97] successfully prepared crack-free 50 wt.% TiC-Ti6Al4V composite gradient coatings using a similar approach.

In summary, the buffer layer serves as a buffer to mitigate cracking issues in composite coatings with a low concentration of hard-phase particles. On the other hand, step coatings are employed to address cracking problems in composite coatings with a high concentration of hard-phase particles. For instance, when the ceramic addition exceeds 50 wt.%, a multilayer step fusion coating should be selected to alleviate stress between the coatings and reduce the susceptibility to cracking.

3.2. *Optimization of Cladding Process Parameters*

3.2.1. Main Process Parameters

The primary process parameters of laser cladding encompass laser power (P), scanning path, scanning speed (V), powder feed (F), powder thickness (H), spot diameter (D), overlap rate (O), and more. Among these parameters, laser power, scanning speed, and spot diameter are interconnected and have a comparable impact on the cracking behavior of the coating. To explore the relationship between laser power (P), scanning speed (V), spot diameter (D), and the coating's cracking behavior, the concept of laser energy density E was introduced [98]:

$$E = \frac{P}{DV} \tag{4}$$

From Formula (4), it can be seen that E is positively correlated with P and E is negatively correlated with D and V. The process parameters of common cladding material systems and their effects on the cracking behavior are shown in Table 2.

Table 2. Relationship between process parameters and crack behavior.

Coating/Substrate	Process Parameters	Influence Law
Ni60/40Cr [99]	*P*: 1000–1800 W, *V*: 20–25 mm/s, F: 80 r/min, *D*: 4 mm, O: 50%, and defocus: 0–4 mm	*P*: 1000–1800 W, the crack rate decreased from 0.2 to 0.11 N/mm. Defocus: 0–4 mm, the crack rate increased from 0.1 to 0.23 N/mm.
FeCoCrNi/ TC4 [63]	*P*: 800–1050 W, V:8–12 mm/s, H: 0.5–1 mm, and *D*: 2 mm	When the heat input and H increased, the cracking first decreased and then increased. When H was 0.75 mm, the crack rate was the lowest. The influence of *P* and *V* on crack rate was much greater than that of H.
TiC/mild Steel [100]	*P*: 1000–2800 W, V: 4 mm, and *D*: 0.5 mm	Under four laser powers, the coating had no cracks or pores, the heat input was large enough, and the coating had no cracks.
WC-Co/ Cr-Mo-V [101]	*P*: 600–1000 W, F: 4–30 g/min, *D*: 2 mm, and nitrogen and argon gas protection atmosphere	When F was higher than 5 g/min, the coating was discontinuous and had defects; under nitrogen protection, all samples had cracks and pores; and under argon protection, all samples had no cracks.
7055AA/ 2024AA [102]	*P*: 1200–2000 W, *V*: 10–30 mm/s, *D*: 2 mm, linear pattern, and two types of arc pattern cladding	The larger *P* was, the smaller *V* was, and the greater residual compressive stress was; the residual stress in linear mode was minimized, and the optimal solution was obtained when the cladding angle was 0°.
StelliteX-40/ GH4133 [103]	*P*: 1000–3000 W, *V*: 5–8 mm/s, H: 0.6–1.2 mm, 750 °C, and 16 h annealing treatment	*P*: 1000 W, the coating was not fused; *P*: 3000 W, reheat cracks appeared; H > 1.1 mm, the coating was not melted; and after treatment, only 3000 W coating had intergranular reheat cracks.
Inconel 690/ Inconel600 [104]	*P*: 2000–3000 W, *V*: 12–16 mm/s, *D*: 5 mm, and 25–300 °C preheating	*P* was too small, or F was too large, leading to poor coating quality; increasing *P*, decreasing F, and preheating could reduce defects.
Ni60/45 Steel [105]	*P*: 1600 W, V: 5–11 mm/s, F: 15–21 g/min, O: 45–60%, and defocus: 16 mm	When F increased, the crack rate increased; when overlap rate increased, the crack rate increased; the larger the *V*, the greater the crack rate.

Here are the revised and retouched paragraphs regarding the recommendations for improving coating cracks:

(1) Increase the laser power appropriately, and reduce the spot diameter and scanning speed. When the laser energy density (E) is low, the molten coating powder may not melt sufficiently, leading to an increased rate of coating cracking. Increasing the laser energy density (E) ensures sufficient melting of the metal in the molten pool, reducing the susceptibility of the coating to cracking [106,107];

(2) Adjust the thickness of powder laying and the amount of powder feeding appropriately. Increasing the thickness of powder laying and the amount of powder feeding results in a larger amount of metal being melted. If the laser energy is insufficient to melt all the powder, it can lead to impurities, pores, coarse coating structure, and, ultimately, cracks. However, reducing the amount of powder feeding or powder thickness too much may result in overburning, thermal deformation of the substrate, and reduced coating quality [105];

(3) Set the overlap rate reasonably. Excessive overlap can cause heat buildup, leading to increased thermal stress and higher susceptibility of the coating to cracking. Conversely, a very low overoverlap rate can result in a discontinuous coating surface and the formation of defects. To ensure high-quality coating formation, the overoverlap rate is generally chosen to be within the range of 30% to 60% [108].

Additionally, the scanning path employed can also impact coating stresses, consequently affecting the propensity for coating cracking. Saboori et al. [109] analyzed the effect of four main scanning strategies (raster, bi-directional, offset, and fractal) on the residual stresses in the final coating. They observed different residual stresses in coatings produced with different scanning paths. For instance, the use of the offset-out scanning strategy reduced the coating residual stresses to one-third of those produced by the bi-directional scanning strategy [110].

3.2.2. Other Process Parameters

In addition to the primary process parameters such as laser power, scanning speed, and powder feed, other factors such as laser parameters (pulse frequency and pulse width) and the external environment (protective atmosphere and incidence angle) can also impact the occurrence of coating cracks.

Decreasing the pulse frequency and increasing the pulse width have been found to reduce coating crack defects. Javid et al. [111] observed a decrease in the number of coating cracks as the pulse width increased. This is attributed to the fact that a wider pulse width results in a lower cooling rate, increased dilution rate, decreased residual stresses, enhanced coating toughness, and reduced susceptibility to cracking. Ali Khorram et al. [112] investigated the influence of parameters such as pulse width and laser frequency on the occurrence of solidification cracking in coatings. They discovered that high dilution rates and coating thickness cause the formation of longer solidification cracks, while crack-free coatings can be achieved at lower dilution rates (28%). Furthermore, utilizing a lower laser frequency can lower the dilution rate, enabling the preparation of crack-free coatings [29].

Although most studies have focused on laser melting coatings on flat surfaces, the investigation of melting conditions on tilted substrates is still in the exploratory stage. Bowen Shi et al. [113] examined the effect of substrate tilt angle on the cracking behavior of high-hardness Ni60A coatings and observed that the cracking rate increased 3.5 times for up-lap coatings and 4 times for down-lap coatings as the tilt angle increased from 0° to 60°. The cracking mode observed was a quasi-dissociative fracture. Therefore, choosing an upward lap configuration can minimize the shielding and reflection of the laser by the molten coating, leading to increased laser utilization, enhanced melt pool mobility, and reduced coating cracking. Additionally, Yamaguchi et al. [114] discovered that reducing the oxygen concentration can effectively prevent defects such as porosity and decarburization resulting from the carbon–oxygen reaction in the coating. Therefore, using an inert gas such as argon as a protective gas can help reduce coating defects such as porosity and cracks.

3.3. Increasing Heat Treatment

Residual stresses in coatings can be mitigated by employing suitable heat treatment techniques during the laser fusion coating process. Commonly used heat treatments include the preheating pretreatment, slow cooling post-treatment, and laser remelting.

Preheating the substrate can reduce the temperature gradient between the coating and the substrate, thereby lowering thermal stress and decreasing the susceptibility of the coating to cracking (Figure 5a). However, excessive preheating can lead to increased crack length, necessitating careful control of the preheating temperature [29]. Ding et al. [115] conducted experiments using different preheating temperatures on Q460E substrates with 12CrNi2 coatings and developed a finite element residual stress analysis model. Their findings demonstrated that preheating effectively reduces residual stresses in the coatings. Bidron et al. [116] conducted repair studies on CM-247LC high-temperature alloy and determined that using low energy input and an induction preheating temperature close to 1100 °C was highly effective in preventing specimen cracking. Moreover, to address various concerns, Zhou et al. [117] proposed the laser induction hybrid rapid cladding (LIHRC) technique, which solved the problem of substrate damage caused by preheating. This technique not only improved cladding efficiency but also resulted in crack-free composite coatings.

Slow cooling post-treatment mainly involves adjusting the microstructure of the coating and slowly releasing residual stress in the coating (Figure 5b). Li et al. [118] found that the fracture toughness of coatings increased from 2.77 to 3.80 MPa·$m^{1/2}$ at 1 and 2 h of post-treatment with heat retention, thereby reducing the susceptibility to cracking. In another study, Chen et al. [119] investigated the effect of post-melt stress relief treatment on residual stresses in laser melting of AISI P20 tool steel on pre-hardened deformed P20 substrates. They observed that slow cooling post-treatment altered the volume fraction of residual austenite in the coating and significantly reduced residual stresses, consequently reducing the coating's susceptibility to cracking. These findings align with the results reported by Chen et al. [120] and Telasang et al. [121]. Dai et al. [78] found that slow cooling post-treatment was not as effective as substrate pre-treatment in eliminating cracks, as some cracks had already formed during the cooling process and could not be eliminated.

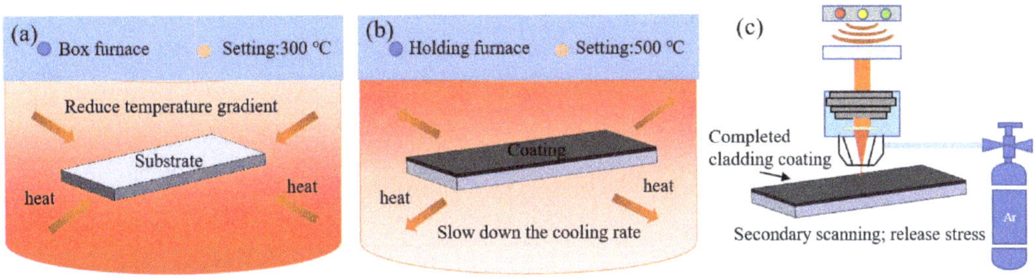

Figure 5. Heat treatments: (**a**) preheating treatment; (**b**) slow cooling post-treatment; (**c**) laser remelting.

Laser remelting is a secondary scanning of the laser cladding coating, which forms a molten pool on the surface of the coating again, thereby releasing coating stress (Figure 5c). Wang et al. [122] prepared iron-based amorphous coatings on H13 steel and observed a significant reduction in coating cracks after laser remelting. Similarly, Lu et al. [123] prepared $Fe_{37.5}Cr_{27.5}C_{12}B_{13}Mo_{10}$ amorphous coatings and found that directly fused coatings exhibited significant cracks, whereas the coatings showed minimal cracks after laser remelting.

In summary, pretreatment is the most effective method for inhibiting coating cracking as it directly reduces thermal stresses. Slow cooling post-treatment functions as a thermal insulation and gradual cooling process to release residual stresses over time. Laser remelting creates a new melt pool on the substrate surface, filling some existing cracks and releasing residual stresses from previous scans. Among the three mentioned techniques, the preheating pretreatment is the most effective in preventing coating cracking. This is because the preheating pretreatment directly reduces thermal stresses. In contrast, post-treatment cannot eliminate existing cracks, and laser remelting may introduce new residual stresses.

3.4. Application of Auxiliary Field

Coating materials and laser coating process parameters have limited direct influence on changes occurring in the melt pool. They can only be adjusted based on process requirements. However, off-field auxiliary technologies have emerged as a popular research topic in recent years to address the cracking problem in laser-clad coatings. These technologies directly impact the energy and mass transfer of the melt pool without physically contacting the coating. They improve the uniformity of the coating structure, reduce elemental segregation, and prevent local stress concentration, thereby slowing down or eliminating defects such as cracks and holes in the coating [85,124–126]. Some commonly studied auxiliary fields include friction stirring, electromagnetic fields, mechanical vibration, and compound fields (Table 3).

Table 3. Effect and mechanism of different auxiliary fields on coating cracks.

Auxiliary Field	Coating/Substrate	Effect	Function Mechanism
Magnetic Field [127]	Co-based/42CrMo	The cracks and pores are reduced; the magnetic field intensity is 20 mT, the magnetostrictive effect is maximum	Magnetostrictive effect can effectively reduce the coefficient of thermal expansion and elastic modulus of cladding layer, and reduce element segregation
Ultrasonic Vibrations [128]	316L/ASTM 1045	Refine the microstructure of the cladding layer, effectively reducing micro defects such as pores and cracks	Vibration causes tissue fragmentation and secondary dendrite fracture, and changes the orientation of the crystal from (220) to (111), resulting in a dense structure

Table 3. *Cont.*

Auxiliary Field	Coating/Substrate	Effect	Function Mechanism
Friction Stir [125]	Ni-Cr-Fe/45 steel	Refines grain size, obtains nanostructured coatings, and eliminates coating cracks	Rapid stirring increases friction heat and strain rate, which is beneficial for toughening and plasticizing
Electromagnetic Field [129]	Ni60/pure iron	The toughness of the coating is improved, and the number of cracks is reduced	Electromagnetic stirring causes Cr_7C_3 to fracture into an independent rod-shaped structure, improving the nucleation rate
Ultrasonic–Electromagnetic Field [130]	NiCrBSi/42CrMo	Refines organization, reduces element segregation, and significantly reduces coating defects	Ultrasonic flow and electromagnetic force stirring cause the dendrites in the cladding layer to be shattered

The electromagnetic field, for instance, can alter the trajectory of charged particles within the melt pool. By using an alternating electromagnetic field, the melt pool can be agitated, and the grains can be crushed. Studies have demonstrated that the magnetic field refines the microstructure of the coating, and magnetostriction significantly reduces the coating's elastic modulus and enhances its elastic recovery. This improvement in crack resistance was observed by Wang et al. [131], who utilized a steady-state magnetic field to assist in melting a nickel-based alloy coating on a pure iron substrate. Without the auxiliary field, five cracks were found on the coating's surface, but as the magnetic induction intensity increased, the number of cracks gradually decreased. When the magnetic induction intensity reached 0.2 T, only two cracks remained.

Ultrasonic vibration-assisted techniques employ ultrasonic cavitation and acoustic flow reinforcement to generate high-frequency and low-amplitude vibrations. These vibrations agitate the melt pool, ensuring a more uniform distribution of elements and stress fields. Zhang et al. [132] prepared a ceramic particle-reinforced iron-based composite coating using ultrasonic vibration assistance and found that the acoustic flow caused by ultrasonic vibration accelerated the melt pool flow, improved heat and mass transfer, and facilitated the formation of nucleation points for ceramic particles. As ultrasonic power increased, the ceramic particles in the coating became finer and more uniform. However, further increasing the ultrasonic power resulted in the agglomeration of fine ceramic particles, increasing the brittleness of the coating and leading to cracking.

Friction stirring involves the application of frictional heat and the coupling effect of plastic deformation on the molten pool, leading to grain refinement and uniform distribution of carbides within the coating. This process enhances the toughness of the coating. Niu et al. [133] employed friction-stirring thermal coupling to eliminate cracks on the top surface of a Ni-16Cr-8Fe alloy coating. They successfully transformed reticulated carbides into nanoparticles dispersed throughout the substrate austenite, thereby improving grain toughness and refinement. Alibadi et al. [134] discovered that the friction stirring-assisted technique outperformed the furnace remelting technique. It not only eliminated defects such as cracks and porosity in the coating but also significantly increased the hardness and wear resistance of the coating.

When it comes to coatings with low ceramic content, a single auxiliary field exhibits a good crack suppression effect. However, for reinforced coatings with high ceramic content, the crack improvement effect is limited. Consequently, researchers have explored composite fields to address this issue. Zhang et al. [135] introduced the ultrasonic–electromagnetic mixed field (UEM) into laser melting (LC) to prepare Inconel 718 + 60 wt.% WC metal substrate composite (MMC) coatings on Inconel 718 substrates. Coatings without assistance exhibited longitudinal and transverse cracks. However, the presence of a single ultrasonic field caused cavitation and a nonlinear effect of acoustic flow, resulting in a more uniform temperature distribution in the melt pool and a reduced temperature gradient. The single UEM field generated two types of Lorentz force in the melt pool, enhancing Marangoni convection and reducing the temperature gradient, leading to a uniform temperature distribution. The combined effect of acoustic and Lorentz forces assisted by mixed UEM fields reduced the drag force of WC in the melt pool [130,136]. Consequently, a larger stirring effect occurred in the melt pool, significantly reducing temperature inhomogeneity and completely suppressing cracks.

In summary, magnetic field assistance exploits magnetostrictive effects, ultrasonic vibration fields utilize acoustic flow reinforcement, electromagnetic fields employ Lorentzian magnetism for stirring, and friction stirring often involves thermal coupling. While a single auxiliary field can eliminate defects such as cracks and holes in general ceramic composite coatings, a composite field is more effective for reinforced coatings with high ceramic content. Both single-assisted and composite fields primarily aim to stir the melt pool, achieving homogeneous tissue and refined grains. As a result, the toughness of the coating improves, thereby reducing its susceptibility to cracks. Off-site auxiliary technology is of great significance for improving the uniformity of microstructure in coatings, refining grain size, reducing residual stress in coatings, and suppressing coating cracks and pore defects in the future.

3.5. Numerical Analysis

Real-time monitoring of the intricate melting, cooling, and solidification processes within the melt pool during laser cladding coating is challenging through experimental means. However, numerical simulation offers a viable solution by simulating the coupled behavior of temperature, stress, and flow fields. This simulation provides valuable insights for optimizing laser cladding process parameters and predicting the expansion of cracks. By employing numerical simulations, researchers gain a comprehensive understanding of the complex phenomena occurring within the melt pool, enabling them to refine and improve the laser cladding process. Additionally, these simulations aid in predicting and mitigating crack formation and expansion, leading to more reliable and efficient coating outcomes.

3.5.1. Melt Pool and Physical Field Model

Regarding the melt pool model, Kumar et al. [137] developed a three-dimensional Marangoni–Rayleigh–Benard convection model to investigate the impact of process parameters, such as power and velocity, on various aspects of the melt pool. Their study examined the geometry of the melt pool, dilution level, melt pool temperature, and the microstructure formation and scale along the solidification trajectory. The results revealed that enhanced uniformity in melt pool flow reduced porosity, contributing to improved outcomes. Similarly, Chew et al. [138] employed a three-dimensional finite element model to simulate the residual stresses induced by laser coating AISI 4340 powders on a similar substrate material. They also proposed a laser power decay model for the laser–powder interaction zone during coaxial powder feeding. Furthermore, researchers such as Aucott et al. [139], using high-energy synchrotron micrographs, observed the formation and flow state of the melt pool during laser cladding, investigating the temperature-dependent alloy surface tension coefficient to optimize the laser additive manufacturing process.

In terms of the physical field model, Yao et al. [140] confirmed the significant advantages of multi-dimensional high-frequency vibration in laser cladding and provided a basis for revealing the potential mechanism of multi-dimensional vibration on rapid melting and solidification. Figure 6a,b in their work displayed the temperature field and flow field of 2D vibration at 0.45 s. Tian et al. [141] employed the stress-induced solid phase transformation (SSPT) technique to mitigate residual stresses in coatings. They developed a three-dimensional thermodynamic finite element model to analyze stress-induced solid phase transformation and found that SSPT significantly reduced residual stresses in both the transverse and longitudinal directions (Figure 6c). By inducing the transformation of austenite to martensite through melting, the residual stresses were released, resulting in low-stress coatings. This approach offered a novel strategy for crack elimination. In another study, Li et al. [142] established a multi-field coupled three-dimensional numerical model of IN625 powder laser disc-clad ductile iron. Their comprehensive model revealed the transient evolution of temperature, flow, morphology, and stress fields within the clad layer.

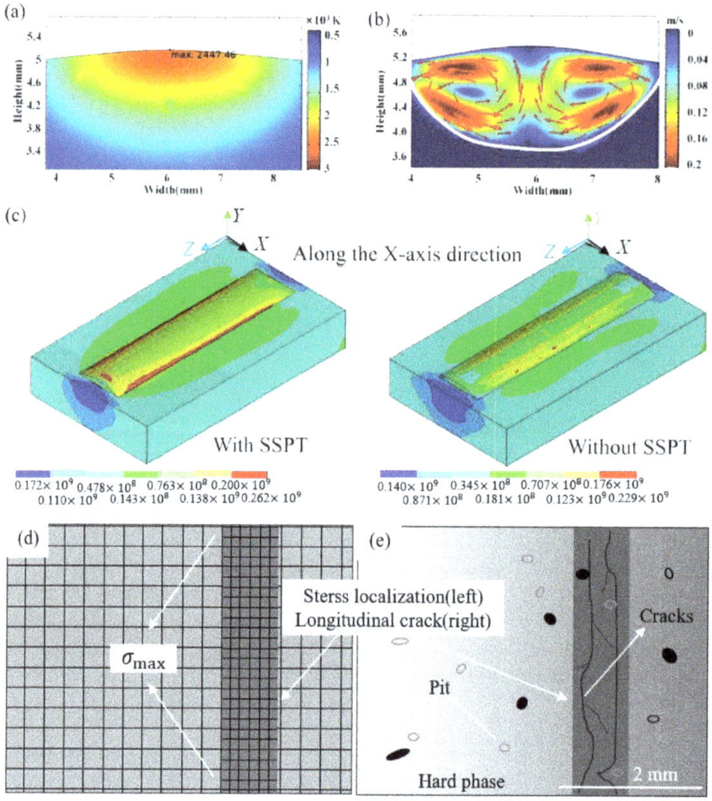

Figure 6. Physical field model and crack prediction. (**a–c**): physical field model (**a**) temperature field with 2D vibration; (**b**) flow field with 2D vibration [140]; (**c**) stress field [141]; and (**d**,**e**): crack prediction. (**d**) Single channel stress distribution; (**e**) actual distribution of coating cracks.

3.5.2. Process Parameters and Crack Prediction Model

When it comes to simulating laser cladding process parameters, Qiao et al. [143] found through numerical simulation that as the overlap rate increases, the heat accumulation and corresponding maximum residual stress tend to be located at the center of the melted coating, at which point the coating has high crack sensitivity. Yu et al. [105] developed a neural network prediction model that correlated the overoverlap rate, powder feed, scanning speed, and coating crack density. They determined that the sensitivity of coating cracks followed the following order: overoverlap rate > powder feed > scanning speed. Song et al. [102] introduced a dual ellipsoidal heat source-based and sequential-coupled thermal force analysis (SCTMA) method to analyze the distribution of temperature and residual stress fields under different laser power, scanning speed, and laser melting parameters. Their findings indicated that the specimens exhibited minimal residual stresses and the best crack resistance when the laser power was set at 1400 W and the scanning speed was 10 mm/s. The simulated laser melting process parameters can provide the theoretical optimal parameters, save the experimental time for optimizing the process parameters, and facilitate the subsequent theoretical support for the crack prediction model. Regarding the prediction of coating cracks, Li et al. [144] found by ANSYS finite element simulation that a high stress concentration was generated between the coating and the substrate, which increased the susceptibility to crack formation, and the distribution of transverse residual stress test was basically consistent with the actual crack orientation. Krzyzanowski et al. [145] conducted numerical simulations to analyze the transient fields of thermal and strain–stress during the melting process. The predicted results demonstrated that the equivalent stress in the marked region on the top surface of the single-pass melting layer exceeded the tensile strength limit of the melted material, resulting in the generation of cracks in that region. This observation aligned with the location of longitudinal cracks observed experimentally (Figure 6d,e). Li et al. [146] proposed a method for identifying the state of the cladding layer and detecting cracks during laser cladding. By monitoring the acoustic emission signal during the laser melting process, the presence of cracks in the coating can be determined, along with identifying their location. Cracked clad layers exhibit high-amplitude and high-energy impact signals, whereas the acoustic emission signal collected from uncracked clad layers is more moderate. The crack prediction model is mainly used to obtain the location of cracks by simulating the stress magnitude and distribution, and to identify the cracks in the coating by detecting the acoustic signal when cracks are generated to determine whether cracks exist in the coating.

In summary, numerous numerical simulations of the laser cladding process have been witnessed in recent years. Scholars have analyzed the flow of the melt pool, as well as the temperature, stress, and flow fields during cladding, while investigating the behavior of cracking under different process parameters. Additionally, the proposed prediction models for coating cracks have facilitated the detection and identification of cracks in laser melting coatings, presenting a significant advancement in crack exploration within laser melting coatings.

3.6. Other Methods

To address the issue of cracks in laser cladding coatings, researchers have proposed innovative methods that go beyond numerical simulations and optimization processes. These approaches mainly involve the development of the exploration of novel preparation processes and new structural materials.

In terms of novel preparation processes, Cen et al. [147] proposed a method for suppressing solidification cracks in Al-Cu-Mg alloy during laser melting using high-frequency beam oscillation (Figure 7a). This method effectively reduced solidification cracks, induced a columnar to equiaxed crystal transformation, and minimized coating cracks without requiring additional auxiliary measures. Yang et al. [148] investigated the forming properties and solidification behavior of a high-reflectivity CuCrZr alloy coating on an AlSi7Mg substrate under infrared (IR), blue, and hybrid lasers. Materials with high crack sensitivity exhibited severe cracking and spheroidization tendencies when melted with IR and blue lasers, respectively. However, the use of a hybrid laser source (960 W blue laser + low-power IR laser of 100 and 200 W) successfully produced crack-free coatings (Figure 7b). Zhang et al. [149] implemented a superposition method where two uniform rectangular spots with different sizes and energy densities were used, resulting in the formation of convex spots (Figure 7c). The thermal-force coupling analysis of the McrAlY coating melted with uniform rectangular and convex beam spot processes demonstrated that the convex spot provided preheating and slow cooling effects, reduced the temperature gradient between the substrate and the melted layer, decreased residual stress, and effectively mitigated cracking tendencies in the melted layer.

In terms of novel structural materials, Feng et al. [150] successfully prepared a crack-free Fe-C-B-W-Cr-Nb-V-Si-Mn coating with an ultrafine eutectic structure consisting of shaped nanostructured dendrites. This coating exhibits the high toughness characteristic of nanomaterials without agglomeration. The small microstructure length of the eutectic structure matches the critical crack size, enabling the possibility of crack deflection at the phase boundary. Another study by Wang et al. [151] involved the preparation of a core–shell-structured composite powder consisting of 420 stainless steel powder and nano WC-Co particles using high-energy ball milling. The composite powder exhibited excellent sphericity and fluidity, making it highly suitable for laser cladding (Figure 6e). Consequently, the application of this core–shell composite powder resulted in the formation of dense coatings with uniform hardness distribution and no cracks.

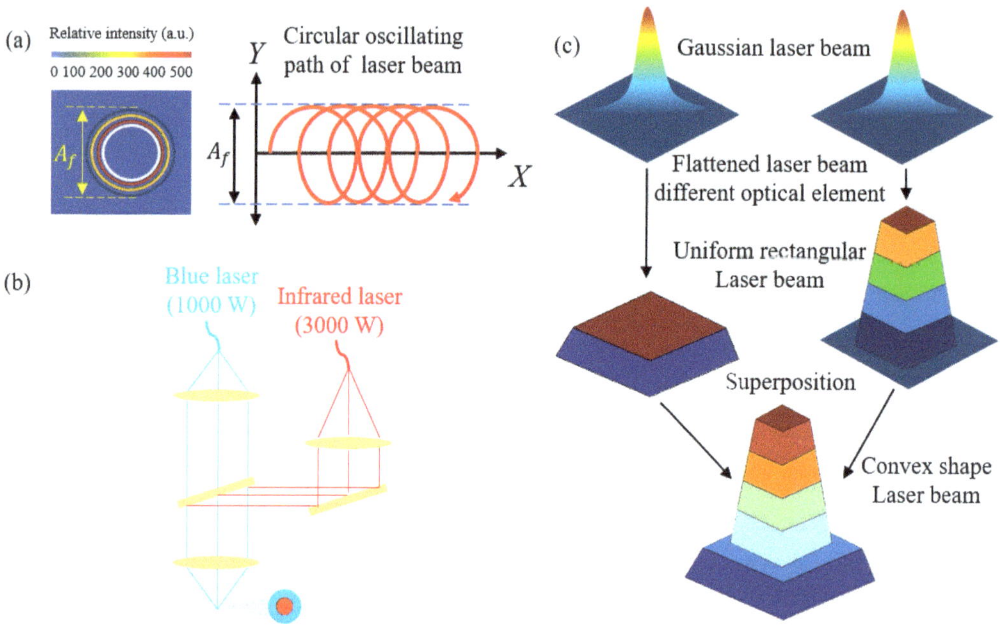

Figure 7. New preparation processes. (**a**) Energy distribution of the circular oscillating laser beam; (**b**) the laser power combination mode; and (**c**) schematic diagram of convex light spots formed by overlapping rectangular light spots.

In summary, researchers have made significant advancements by exploring innovative preparation processes and new structural materials to control cracks in laser cladding coatings. These approaches offer promising strategies for achieving crack-free coatings with improved properties.

3.7. Cracks Control Methods Comparison and Evaluation

In view of the problem that the variety of coating crack control methods is complicated and the practicality is difficult to evaluate, this paper adds a comparative analysis of various control methods, evaluating and analyzing them in terms of control effects, economic benefits, applicability, and other aspects. Table 4 summarizes the practical effects, advantages, and disadvantages of six crack control methods in order to provide an effective reference value for laser cladding coating crack control.

As shown in Table 4, among the six coating crack control methods, the addition of rare earth oxides, the preheating treatment, compound field assistance, new preparation processes, and new structural materials are the most effective for the control of laser melting coating cracks, while post-treatment and laser remelting are generally effective. Numerical analysis can study the melt pool state and thus predict the cracking behavior, where controlling the hard-phase content and optimizing the process parameters are the most widely used. Application of auxiliary fields, numerical analysis, and development of new structural materials and processes are the hotspots of current research with good development prospects.

Table 4. Practical effects, advantages, and disadvantages of the six crack control methods.

Control Methods	Specific Types	Practical Effects	Advantages	Disadvantages
Preferred Cladding Materials	Control of hard-phase contents in the layer	2	Easy to use, wide range of applications	Difficult to eliminate cracks, not applicable without hard phase coating
	Preferred hard-phase particle types	2	Better research prospects	Not applicable without hard-phase coating
	Adding alloying elements	1	Easy to use	Few types of alloy elements and may affect coating properties
	Adding rare earth oxides	0	Wide application range and improving coating properties	High cost, fewer types of rare earth oxides
	Adding transition layer	2	Can reduce coating cracks with high hard-phase content	Small scope of application
Optimization of Cladding Process Parameters	Main process parameters	1	The widest range of applications	Difficult to eliminate cracks in coatings with high hard-phase content
	Other process parameters	2	Better research prospects	High requirements for experimental equipment
Increasing Heat Treatment	Preheat treatment	0	Wide application range	Longer waiting time
	Post-processing	3	Easy to use	The effect is average
	Laser remelting	3	Improving coating properties	Additional costs required
Application of Auxiliary Field	Magnetic field	1	Highly promising research improving coating properties	High requirements for experimental equipment, still in the experimental stage
	Ultrasonic vibrations	1		
	Friction stirring	1		
	Electromagnetic field	1		
	Ultrasonic–electromagnetic field	0	Eliminates cracks in coatings with high hard-phase content	Complex coupling process, extremely demanding equipment
Numerical Analysis	Melt pool and Physical field model	Provide stress distribution diagram	Exploring the melt pool and providing stress distribution diagram	Model complexity and accuracy need to be improved
	Process Parameters and Crack Prediction Model	Predicts optimal processes and cracks	Highly promising research, reducing experiment cost	Longer simulation time and more difficult operation
Other Methods	Novel preparation processes	0	Extremely high research prospect	The research is difficult and still in the exploration stage
	Novel structural materials	0		

Notes: (0) No cracks, (1) almost no cracks, (2) small amounts of cracks, and (3) some cracks remain.

4. Conclusions and Outlook

This paper provides a comprehensive review of crack types, formation mechanisms, and control methods in laser cladding coatings. Additionally, it introduces new approaches involving innovative preparation processes and novel structural materials for crack control in laser cladding coatings.

Cracks in laser cladding coatings primarily arise from residual stresses. The uneven distribution of temperature, stress, and flow fields in the melt pool, along with the temperature gradient and the difference in thermal expansion coefficients and elastic moduli between the molten material and the substrate, make complete elimination of residual stress challenging. Therefore, current research focuses on regulating thermal stress, organizational stress, and restraint stress, reducing individual or interactive stress effects, and inhibiting or eliminating coating cracks during the laser melting process.

To mitigate residual stresses, various control methods are employed, including the selective choice of cladding layer materials, optimization of process parameters, implementation of heat treatment, and application of auxiliary fields. Optimal selection of cladding materials involves choosing hard-phase particles with suitable size, shape, and structure, as well as adding appropriate proportions of alloying elements or rare earth oxides. Process parameter optimization, such as selecting suitable laser cladding parameters and scanning strategies, is crucial. Heat treatment methods such as preheating, slow cooling post-treatment, and laser remelting are employed. The application of auxiliary fields, such as magnetic fields, ultrasonic vibration, electromagnetic fields, friction stirring, and multi-field compounding, is also significant. Furthermore, numerical simulation of the laser cladding process and the identification and prediction of coating cracks offer valuable perspectives for experimental analysis.

New preparation processes and new structural materials have become a new direction for crack control in laser cladding coatings in recent years. These methods include the use of high-frequency beam oscillation, hybrid laser sources, spot superposition technology, ultra-fine eutectic organization, and core–shell-structured composite powders. These techniques contribute to denser coating structures, reduced residual stresses, and effectively slowing down crack formation tendencies.

However, the current research is mainly focused on the experiments of material proportioning and cladding process parameters as well as the establishment of basic mathematical models. Further control of cracks in laser cladding coatings should be carried out from the following aspects:

(1) Given that the forming mechanism of the melt pool and stress distribution's impact on coating cracks remains unexplored, establishing a comprehensive mathematical model to simulate the entire laser cladding process is necessary. This would provide a theoretical foundation for analyzing crack control in the cladding layer and avoid blind trial and error.

(2) The selection of hard-phase particles in laser cladding coatings lacks sufficient basis, necessitating the development of a standardized table to guide the particle selection process and prevent cracks caused by the variation in actual stress due to different particles.

(3) While the addition of rare earth oxides greatly enhances coating quality, existing additives are limited in variety, expensive, and not suitable for large-scale applications. Thus, new low-cost, high-performance additives need to be developed.

(4) The development of real-time monitoring systems that utilize imaging, temperature sensing, spectral analysis, and sound detection is essential. This would enable real-time regulation of the melt pool temperature and power fluctuations, thereby reducing coating defects and monitoring crack behavior.

(5) As new preparation processes and structural materials significantly improve coating quality and reduce stress, the further exploration and integration of crack control technology with composite preparation processes are crucial to driving the industrial application of laser cladding technology.

Author Contributions: Conceptualization, M.L. and K.H.; writing—review and editing, X.Y.; supervision, X.Y.; project administration. All authors have read and agreed to the published version of the manuscript.

Funding: This research received no external funding.

Institutional Review Board Statement: Not applicable.

Informed Consent Statement: Not applicable.

Data Availability Statement: Not applicable.

Conflicts of Interest: The authors declare no conflict of interest.

References

1. Zhu, L.; Xue, P.; Lan, Q.; Meng, G.; Ren, Y.; Yang, Z.; Xu, P.; Liu, Z. Recent Research and Development Status of Laser Cladding: A Review. *Opt. Laser Technol.* **2021**, *138*, 106915. [CrossRef]
2. He, B.; Zhang, L.; Zhu, Q.; Wang, J.; Yun, X.; Luo, J.; Chen, Z. Effect of Solution Treated 316L Layer Fabricated by Laser Cladding on Wear and Corrosive Wear Resistance. *Opt. Laser Technol.* **2020**, *121*, 105788. [CrossRef]
3. Wang, Q.-Y.; Zhang, Y.-F.; Bai, S.-L.; Liu, Z.-D. Microstructures, Mechanical Properties and Corrosion Resistance of Hastelloy C22 Coating Produced by Laser Cladding. *J. Alloys Compd.* **2013**, *553*, 253–258. [CrossRef]
4. Lu, X.-L.; Liu, X.-B.; Yu, P.-C.; Qiao, S.-J.; Zhai, Y.-J.; Wang, M.-D.; Chen, Y.; Xu, D. Synthesis and Characterization of Ni60-HBN High Temperature Self-Lubricating Anti-Wear Composite Coatings on Ti6Al4V Alloy by Laser Cladding. *Opt. Laser Technol.* **2016**, *78*, 87–94. [CrossRef]
5. Wu, C.L.; Zhang, S.; Zhang, C.H.; Zhang, J.B.; Liu, Y.; Chen, J. Effects of SiC Content on Phase Evolution and Corrosion Behavior of SiC-Reinforced 316L Stainless Steel Matrix Composites by Laser Melting Deposition. *Opt. Laser Technol.* **2019**, *115*, 134–139. [CrossRef]

6. Zhang, P.; Liu, X.; Yan, H. Phase Composition, Microstructure Evolution and Wear Behavior of Ni-Mn-Si Coatings on Copper by Laser Cladding. *Surf. Coat. Technol.* **2017**, *332*, 504–510. [CrossRef]
7. Siddiqui, A.A.; Dubey, A.K. Recent Trends in Laser Cladding and Surface Alloying. *Opt. Laser Technol.* **2021**, *134*, 106619. [CrossRef]
8. Chen, L.; Yu, T.; Chen, X.; Zhao, Y.; Guan, C. Process Optimization, Microstructure and Microhardness of Coaxial Laser Cladding TiC Reinforced Ni-Based Composite Coatings. *Opt. Laser Technol.* **2022**, *152*, 108129. [CrossRef]
9. Lu, J.Z.; Xue, K.N.; Lu, H.F.; Xing, F.; Luo, K.Y. Laser Shock Wave-Induced Wear Property Improvement and Formation Mechanism of Laser Cladding Ni25 Coating on H13 Tool Steel. *J. Mater. Process. Technol.* **2021**, *296*, 117202. [CrossRef]
10. Guo, C.; Chen, J.; Zhou, J.; Zhao, J.; Wang, L.; Yu, Y.; Zhou, H. Effects of WC–Ni Content on Microstructure and Wear Resistance of Laser Cladding Ni-Based Alloys Coating. *Surf. Coat. Technol.* **2012**, *206*, 2064–2071. [CrossRef]
11. Song, J.; Deng, Q.; Chen, C.; Hu, D.; Li, Y. Rebuilding of Metal Components with Laser Cladding Forming. *Appl. Surf. Sci.* **2006**, *252*, 7934–7940. [CrossRef]
12. Lu, M.; McCormick, P.; Zhao, Y.; Fan, Z.; Huang, H. Laser Deposition of Compositionally Graded Titanium Oxide on Ti6Al4V Alloy. *Ceram. Int.* **2018**, *44*, 20851–20861. [CrossRef]
13. Lian, G.; Yao, M.; Zhang, Y.; Huang, X. Analysis and Respond Surface Methodology Modeling on Property and Performance of Two-Dimensional Gradient Material Laser Cladding on Die-Cutting Tool. *Materials* **2018**, *11*, 2052. [CrossRef] [PubMed]
14. Zhang, J.; Song, B.; Wei, Q.; Bourell, D.; Shi, Y. A Review of Selective Laser Melting of Aluminum Alloys: Processing, Microstructure, Property and Developing Trends. *J. Mater. Sci. Technol.* **2019**, *35*, 270–284. [CrossRef]
15. Galy, C.; Le Guen, E.; Lacoste, E.; Arvieu, C. Main Defects Observed in Aluminum Alloy Parts Produced by SLM: From Causes to Consequences. *Addit. Manuf.* **2018**, *22*, 165–175. [CrossRef]
16. Quazi, M.M.; Fazal, M.A.; Haseeb, A.S.M.A.; Yusof, F.; Masjuki, H.H.; Arslan, A. Effect of Rare Earth Elements and Their Oxides on Tribo-Mechanical Performance of Laser Claddings: A Review. *J. Rare Earths* **2016**, *34*, 549–564. [CrossRef]
17. Wang, X.; Lei, L.; Yu, H. A Review on Microstructural Features and Mechanical Properties of Wheels/Rails Cladded by Laser Cladding. *Micromachines* **2021**, *12*, 152. [CrossRef] [PubMed]
18. Hu, Y.; Cong, W. A Review on Laser Deposition-Additive Manufacturing of Ceramics and Ceramic Reinforced Metal Matrix Composites. *Ceram. Int.* **2018**, *44*, 20599–20612. [CrossRef]
19. Fang, Z.-C.; Wu, Z.-L.; Huang, C.-G.; Wu, C.-W. Review on Residual Stress in Selective Laser Melting Additive Manufacturing of Alloy Parts. *Opt. Laser Technol.* **2020**, *129*, 106283. [CrossRef]
20. Sanaei, N.; Fatemi, A. Defects in Additive Manufactured Metals and Their Effect on Fatigue Performance: A State-of-the-Art Review. *Prog. Mater. Sci.* **2021**, *117*, 100724. [CrossRef]
21. Cheng, J.; Xing, Y.; Dong, E.; Zhao, L.; Liu, H.; Chang, T.; Chen, M.; Wang, J.; Lu, J.; Wan, J. An Overview of Laser Metal Deposition for Cladding: Defect Formation Mechanisms, Defect Suppression Methods and Performance Improvements of Laser-Cladded Layers. *Materials* **2022**, *15*, 5522. [CrossRef] [PubMed]
22. Wang, F.; Mao, H.; Zhang, D.; Zhao, X.; Shen, Y. Online Study of Cracks during Laser Cladding Process Based on Acoustic Emission Technique and Finite Element Analysis. *Appl. Surf. Sci.* **2008**, *255*, 3267–3275. [CrossRef]
23. Wang, D.; Liang, E.; Chao, M.; Yuan, B. Investigation on the Microstructure and Cracking Susceptibility of Laser-Clad V_2O_5/NiCrBSiC Alloy Coatings. *Surf. Coat. Technol.* **2008**, *202*, 1371–1378. [CrossRef]
24. Cloots, M.; Uggowitzer, P.J.; Wegener, K. Investigations on the Microstructure and Crack Formation of IN738LC Samples Processed by Selective Laser Melting Using Gaussian and Doughnut Profiles. *Mater. Des.* **2016**, *89*, 770–784. [CrossRef]
25. Eo, D.-R.; Chung, S.-G.; Yang, J.; Cho, W.T.; Park, S.-H.; Cho, J.-W. Surface Modification of High-Mn Steel via Laser-DED: Microstructural Characterization and Hot Crack Susceptibility of Clad Layer. *Mater. Des.* **2022**, *223*, 111188. [CrossRef]
26. Alizadeh-Sh, M.; Marashi, S.P.H.; Ranjbarnodeh, E.; Shoja-Razavi, R. Laser Cladding of Inconel 718 Powder on a Non-Weldable Substrate: Clad Bead Geometry-Solidification Cracking Relationship. *J. Manuf. Process.* **2020**, *56*, 54–62. [CrossRef]
27. Shankar, V.; Gill, T.P.S.; Mannan, S.L.; Terrance, A.L.E.; Sundaresan, S. Relation between Microstructure, Composition, and Hot Cracking in Ti-Stabilized Austenitic Stainless Steel Weldments. *Metall. Mater. Trans. A* **2000**, *31*, 3109–3122. [CrossRef]
28. Chen, Y.; Lu, F.; Zhang, K.; Nie, P.; Elmi Hosseini, S.R.; Feng, K.; Li, Z. Dendritic Microstructure and Hot Cracking of Laser Additive Manufactured Inconel 718 under Improved Base Cooling. *J. Alloys Compd.* **2016**, *670*, 312–321. [CrossRef]
29. Ebrahimzadeh, H.; Farhangi, H.; Mousavi, S.A.A.A. Hot Cracking in Autogenous Welding of 6061-T6 Aluminum Alloy by Rectangular Pulsed Nd: YAG Laser Beam. *Weld. World* **2020**, *64*, 1077–1088. [CrossRef]
30. Zenitani, S.; Hayakawa, N.; Yamamoto, J.; Hiraoka, K.; Morikage, Y.; Kubo, T.; Yasuda, K.; Amano, K. Development of New Low Transformation Temperature Welding Consumable to Prevent Cold Cracking in High Strength Steel Welds. *Sci. Technol. Weld. Join.* **2007**, *12*, 516–522. [CrossRef]
31. Hu, L.H.; Huang, J.; Li, Z.G.; Wu, Y.X. Effects of Preheating Temperature on Cold Cracks, Microstructures and Properties of High Power Laser Hybrid Welded 10Ni3CrMoV Steel. *Mater. Des.* **2011**, *32*, 1931–1939. [CrossRef]
32. Boes, J.; Röttger, A.; Theisen, W. Processing of X65MoCrWV3-2 Cold Work Tool Steel by Laser Powder Bed Fusion. *Steel Res. Int.* **2020**, *91*, 1900445. [CrossRef]
33. Li, K.; Wang, X.; Brodusch, N.; Tu, G. Mitigating Cracking in Laser Powder Bed Fusion of Ti-48Al-2Cr-2Nb via Introducing Massive β Phase. *Mater. Charact.* **2023**, *196*, 112558. [CrossRef]

34. Sadhu, A.; Choudhary, A.; Sarkar, S.; Nair, A.M.; Nayak, P.; Pawar, S.D.; Muvvala, G.; Pal, S.K.; Nath, A.K. A Study on the Influence of Substrate Pre-Heating on Mitigation of Cracks in Direct Metal Laser Deposition of NiCrSiBC-60%WC Ceramic Coating on Inconel 718. *Surf. Coat. Technol.* **2020**, *389*, 125646. [CrossRef]
35. Liu, S.; Sun, Y.; Zhai, P.; Fan, P.; Zhang, Y.; Li, M.; Fang, J.; Wu, R.; Cai, Z. Microstructure and Properties of Nickel-Based Gradient Coatings Prepared Using Cold Spraying Combined with Laser Cladding Methods. *Materials* **2023**, *16*, 1627. [CrossRef]
36. Dmitrieva, A.; Klimova-Korsmik, O.; Gushchina, M.; Korsmik, R.; Zadykyan, G.; Tukov, S. Effect of the Laser Cladding Parameters on the Crack Formation and Microstructure during Nickel Superalloy Gas Turbine Engines Repair. *Metals* **2023**, *13*, 393. [CrossRef]
37. Zhang, Z.; Zhao, Y.; Chen, Y.; Su, Z.; Shan, J.; Wu, A.; Sato, Y.S.; Gu, H.; Tang, X. The Role of the Pulsed-Wave Laser Characteristics on Restraining Hot Cracking in Laser Cladding Non-Weldable Nickel-Based Superalloy. *Mater. Des.* **2021**, *198*, 109346. [CrossRef]
38. Meng, L.; Sheng, P.; Zeng, X. Comparative Studies on the Ni60 Coatings Deposited by Conventional and Induction Heating Assisted Extreme-High-Speed Laser Cladding Technology: Formability, Microstructure and Hardness. *J. Mater. Res. Technol.* **2022**, *16*, 1732–1746. [CrossRef]
39. Ma, Q.; Lu, B.; Zhang, Y.; Wang, Y.; Yan, X.; Liu, M.; Zhao, G. Crack-Free 60 Wt% WC Reinforced FeCoNiCr High-Entropy Alloy Composite Coating Fabricated by Laser Cladding. *Mater. Lett.* **2022**, *324*, 132667. [CrossRef]
40. Wang, K.; Liu, B.; Wang, Q.; Bao, R. Macrostructures and Mechanical Properties in Laser Melting Deposited Titanium Alloy Plate at Different Thickness Positions. *Mater. Sci. Eng. A* **2022**, *832*, 142433. [CrossRef]
41. Wang, K.; Bao, R.; Liu, D.; Yan, C. Plastic Anisotropy of Laser Melting Deposited Ti–5Al–5Mo–5V–1Cr–1Fe Titanium Alloy. *Mater. Sci. Eng. A* **2019**, *746*, 276–289. [CrossRef]
42. Mishra, R.S.; Ma, Z.Y. Friction Stir Welding and Processing. *Mater. Sci. Eng. R Rep.* **2005**, *50*, 1–78. [CrossRef]
43. Wu, P.; Du, H.M.; Chen, X.L.; Li, Z.Q.; Bai, H.L.; Jiang, E.Y. Influence of WC Particle Behavior on the Wear Resistance Properties of Ni–WC Composite Coatings. *Wear* **2004**, *257*, 142–147. [CrossRef]
44. Bussu, G.; Irving, P. The Role of Residual Stress and Heat Affected Zone Properties on Fatigue Crack Propagation in Friction Stir Welded 2024-T351 Aluminium Joints. *Int. J. Fatigue* **2003**, *25*, 77–88. [CrossRef]
45. Zhang, Z.; Kovacevic, R. A Thermo-Mechanical Model for Simulating the Temperature and Stress Distribution during Laser Cladding Process. *Int. J. Adv. Manuf. Technol.* **2019**, *102*, 457–472. [CrossRef]
46. Zhou, S.; Zeng, X.; Hu, Q.; Huang, Y. Analysis of Crack Behavior for Ni-Based WC Composite Coatings by Laser Cladding and Crack-Free Realization. *Appl. Surf. Sci.* **2008**, *255*, 1646–1653. [CrossRef]
47. Ignat, S.; Sallamand, P.; Nichici, A.; Vannes, B.; Grevey, D.; Cicală, E. $MoSi_2$ Laser Cladding—Elaboration, Characterisation and Addition of Non-Stabilized ZrO_2 Powder Particles. *Intermetallics* **2003**, *11*, 931–938. [CrossRef]
48. Shah, K.; Khurshid, H.; ul Haq, I.; Anwar, S.; Shah, S.A. Numerical Modelling of Pulsed and Continuous Wave Direct Laser Deposition of Ti-6Al-4V and Inconel 718. *Int. J. Adv. Manuf. Technol.* **2018**, *95*, 847–860. [CrossRef]
49. Sun, S.; Fu, H.; Chen, S.; Ping, X.; Wang, K.; Guo, X.; Lin, J.; Lei, Y. A Numerical-Experimental Investigation of Heat Distribution, Stress Field and Crack Susceptibility in Ni60A Coatings. *Opt. Laser Technol.* **2019**, *117*, 175–185. [CrossRef]
50. Ramakrishnan, A.; Dinda, G.P. Direct Laser Metal Deposition of Inconel 738. *Mater. Sci. Eng. A* **2019**, *740–741*, 1–13. [CrossRef]
51. Brauner, N.; Shacham, M. Statistical Analysis of Linear and Nonlinear Correlation of the Arrhenius Equation Constants. *Chem. Eng. Process. Process Intensif.* **1997**, *36*, 243–249. [CrossRef]
52. Zhou, Z.; Di, X.; Li, C.; Wu, S. Effect of Restraint Stress on Martensite Transformation in Low Transformation Temperature Weld Metal. *J. Mater. Sci.* **2020**, *55*, 2202–2214. [CrossRef]
53. Bendeich, P.; Alam, N.; Brandt, M.; Carr, D.; Short, K.; Blevins, R.; Curfs, C.; Kirstein, O.; Atkinson, G.; Holden, T.; et al. Residual Stress Measurements in Laser Clad Repaired Low Pressure Turbine Blades for the Power Industry. *Mater. Sci. Eng. A* **2006**, *437*, 70–74. [CrossRef]
54. Kirchlechner, C.; Martinschitz, K.J.; Daniel, R.; Klaus, M.; Genzel, C.; Mitterer, C.; Keckes, J. Residual Stresses and Thermal Fatigue in CrN Hard Coatings Characterized by High-Temperature Synchrotron X-Ray Diffraction. *Thin Solid Films* **2010**, *518*, 2090–2096. [CrossRef]
55. Dixneit, J.; Kromm, A.; Hannemann, A.; Friedersdorf, P.; Kannengiesser, T.; Gibmeier, J. In-Situ Load Analysis in Multi-Run Welding Using LTT Filler Materials. *Weld. World* **2016**, *60*, 1159–1168. [CrossRef]
56. Yu, T.; Deng, Q.; Dong, G.; Yang, J. Effects of Ta on Microstructure and Microhardness of Ni Based Laser Clad Coating. *Appl. Surf. Sci.* **2011**, *257*, 5098–5103. [CrossRef]
57. Zhang, S.H.; Li, M.X.; Cho, T.Y.; Yoon, J.H.; Lee, C.G.; He, Y.Z. Laser Clad Ni-Base Alloy Added Nano- and Micron-Size CeO_2 Composites. *Opt. Laser Technol.* **2008**, *40*, 716–722. [CrossRef]
58. Zheng, B.J.; Chen, X.M.; Lian, J.S. Microstructure and Wear Property of Laser Cladding Al+SiC Powders on AZ91D Magnesium Alloy. *Opt. Lasers Eng.* **2010**, *48*, 526–532. [CrossRef]
59. Xu, G.; Kutsuna, M.; Liu, Z.; Zhang, H. Characteristics of Ni-Based Coating Layer Formed by Laser and Plasma Cladding Processes. *Mater. Sci. Eng. A* **2006**, *417*, 63–72. [CrossRef]
60. Huang, L.; Zhou, J.; Xu, J.; Huo, K.; He, W.; Meng, X.; Huang, S. Microstructure and Wear Resistance of Electromagnetic Field Assisted Multi-Layer Laser Clad Fe901 Coating. *Surf. Coat. Technol.* **2020**, *395*, 125876. [CrossRef]
61. Li, X.P.; Roberts, M.; Liu, Y.J.; Kang, C.W.; Huang, H.; Sercombe, T.B. Effect of Substrate Temperature on the Interface Bond between Support and Substrate during Selective Laser Melting of Al–Ni–Y–Co–La Metallic Glass. *Mater. Des. (1980–2015)* **2015**, *65*, 1–6. [CrossRef]

62. Gan, Z.; Yu, G.; He, X.; Li, S. Surface-Active Element Transport and Its Effect on Liquid Metal Flow in Laser-Assisted Additive Manufacturing. *Int. Commun. Heat Mass Transf.* **2017**, *86*, 206–214. [CrossRef]
63. Gao, Z.; Wang, L.; Wang, Y.; Lyu, F.; Zhan, X. Crack Defects and Formation Mechanism of FeCoCrNi High Entropy Alloy Coating on TC4 Titanium Alloy Prepared by Laser Cladding. *J. Alloys Compd.* **2022**, *903*, 163905. [CrossRef]
64. Ning, Z.; Li, T.; Hui, G.; Mengqi, Z. Effect of Ultra-Fine WC Particles on Microstructural Evolution and Wear Behavior of Ni-Based Nano-CeO2 Coatings Produced by Laser. *Rare Met. Mater. Eng.* **2018**, *47*, 20–25. [CrossRef]
65. Shen, X.; Peng, H.; Xue, Y.; Wang, B.; Su, G.; Zhu, J.; Li, A. Microstructure and Properties of WC/Ni-Based Laser-Clad Coatings with Different WC Content Values. *Materials* **2022**, *15*, 6309. [CrossRef]
66. Shen, X.; He, X.; Gao, L.; Su, G.; Xu, C.; Xu, N. Study on Crack Behavior of Laser Cladding Ceramic-Metal Composite Coating with High Content of WC. *Ceram. Int.* **2022**, *48*, 17460–17470. [CrossRef]
67. Liu, D.; Zhang, S.Q.; Li, A.; Wang, H.M. Microstructure and Tensile Properties of Laser Melting Deposited TiC/TA15 Titanium Matrix Composites. *J. Alloys Compd.* **2009**, *485*, 156–162. [CrossRef]
68. Zhang, P.; Pang, Y.; Yu, M. Effects of WC Particle Types on the Microstructures and Properties of WC-Reinforced Ni60 Composite Coatings Produced by Laser Cladding. *Metals* **2019**, *9*, 583. [CrossRef]
69. Sun, Z.; Ma, Y.; Ponge, D.; Zaefferer, S.; Jägle, E.A.; Gault, B.; Rollett, A.D.; Raabe, D. Thermodynamics-Guided Alloy and Process Design for Additive Manufacturing. *Nat. Commun.* **2022**, *13*, 4361. [CrossRef]
70. Chen, H.; Xu, C.; Chen, J.; Zhao, H.; Zhang, L.; Wang, Z. Microstructure and Phase Transformation of WC/Ni60B Laser Cladding Coatings during Dry Sliding Wear. *Wear* **2008**, *264*, 487–493. [CrossRef]
71. Paul, C.P.; Toyserkani, E.; Khajepour, A.; Corbin, S. Cladding of WC–12 Co on Low Carbon Steel Using a Pulsed Nd:YAG Laser. *Mater. Sci. Eng. A* **2007**, *464*, 170–176. [CrossRef]
72. Wang, K.L.; Zhang, Q.B.; Sun, M.L.; Wei, X.G. Microstructural Characteristics of Laser Clad Coatings with Rare Earth Metal Elements. *J. Mater. Process. Technol.* **2003**, *139*, 448–452. [CrossRef]
73. He, X.; Song, R.G.; Kong, D.J. Effects of TiC on the Microstructure and Properties of TiC/TiAl Composite Coating Prepared by Laser Cladding. *Opt. Laser Technol.* **2019**, *112*, 339–348. [CrossRef]
74. Wang, H.; Zuo, D.; Sun, Y.; Xu, F.; Zhang, D. Microstructure of Nanometer Al_2O_3 Dispersion Strengthened Ni-Based High-Temperature Protective Coatings by Laser Cladding. *Trans. Nonferrous Met. Soc. China* **2009**, *19*, 586–591. [CrossRef]
75. Li, X.; Yang, X.; Yi, D.; Liu, B.; Zhu, J.; Li, J.; Gao, C.; Wang, L. Effects of NbC Content on Microstructural Evolution and Mechanical Properties of Laser Cladded Fe50Mn30Co10Cr10-XNbC Composite Coatings. *Intermetallics* **2021**, *138*, 107309. [CrossRef]
76. Zhang, J.; Lei, J.; Gu, Z.; Tantai, F.; Tian, H.; Han, J.; Fang, Y. Effect of WC-12Co Content on Wear and Electrochemical Corrosion Properties of Ni-Cu/WC-12Co Composite Coatings Deposited by Laser Cladding. *Surf. Coat. Technol.* **2020**, *393*, 125807. [CrossRef]
77. Mazaheri Tehrani, H.; Shoja-Razavi, R.; Erfanmanesh, M.; Hashemi, S.H.; Barekat, M. Evaluation of the Mechanical Properties of WC-Ni Composite Coating on an AISI 321 Steel Substrate. *Opt. Laser Technol.* **2020**, *127*, 106138. [CrossRef]
78. Dai, Q.-L.; Luo, C.; You, F. Crack Restraining Methods and Their Effects on the Microstructures and Properties of Laser Cladded WC/Fe Coatings. *Materials* **2018**, *11*, 2541. [CrossRef]
79. Qi, K.; Yang, Y.; Liang, W.; Jin, K.; Xiong, L. Influence of the Anomalous Elastic Modulus on the Crack Sensitivity and Wear Properties of Laser Cladding under the Effects of a Magnetic Field and Cr Addition. *Surf. Coat. Technol.* **2021**, *423*, 127575. [CrossRef]
80. Sun, S.; Fu, H.; Ping, X.; Guo, X.; Lin, J.; Lei, Y.; Wu, W.; Zhou, J. Effect of CeO_2 Addition on Microstructure and Mechanical Properties of In-Situ (Ti, Nb) C/Ni Coating. *Surf. Coat. Technol.* **2019**, *359*, 300–313. [CrossRef]
81. Shu, D.; Dai, S.; Wang, G.; Si, W.; Xiao, P.; Cui, X.; Chen, X. Influence of CeO_2 Content on WC Morphology and Mechanical Properties of WC/Ni Matrix Composites Coating Prepared by Laser in-Situ Synthesis Method. *J. Mater. Res. Technol.* **2020**, *9*, 11111–11120. [CrossRef]
82. Zhang, M.; Wang, X.H.; Qu, K.L.; Liu, S.S. Effect of Rare Earth Oxide on Microstructure and High Temperature Oxidation Properties of Laser Cladding Coatings on 5CrNiMo Die Steel Substrate. *Opt. Laser Technol.* **2019**, *119*, 105597. [CrossRef]
83. Li, J.; Luo, X.; Li, G.J. Effect of Y_2O_3 on the Sliding Wear Resistance of TiB/TiC-Reinforced Composite Coatings Fabricated by Laser Cladding. *Wear* **2014**, *310*, 72–82. [CrossRef]
84. Li, M.; Han, B.; Wang, Y.; Pu, K. Effects of La_2O_3 on the Microstructure and Property of Laser Cladding Ni-Based Ceramic Coating. *Optik* **2017**, *130*, 1032–1037. [CrossRef]
85. Farahmand, P.; Liu, S.; Zhang, Z.; Kovacevic, R. Laser Cladding Assisted by Induction Heating of Ni–WC Composite Enhanced by Nano-WC and La_2O_3. *Ceram. Int.* **2014**, *40*, 15421–15438. [CrossRef]
86. Tian, Y.S.; Chen, C.Z.; Chen, L.X.; Huo, Q.H. Effect of RE Oxides on the Microstructure of the Coatings Fabricated on Titanium Alloys by Laser Alloying Technique. *Scr. Mater.* **2006**, *54*, 847–852. [CrossRef]
87. Zhang, Z.; Yang, F.; Zhang, H.; Zhang, T.; Wang, H.; Xu, Y.; Ma, Q. Influence of CeO_2 Addition on Forming Quality and Microstructure of TiC -Reinforced CrTi4-Based Laser Cladding Composite Coating. *Mater. Charact.* **2021**, *171*, 110732. [CrossRef]
88. Yuling, G.; Meiping, W.; Xiaojin, M.; Chen, C. Effect of CeO_2 on Crack Sensitivity and Tribological Properties of Ni60A Coatings Prepared by Laser Cladding. *Adv. Mech. Eng.* **2021**, *13*, 168781402110131. [CrossRef]
89. Wang, C.; Gao, Y.; Zeng, Z.; Fu, Y. Effect of Rare-Earth on Friction and Wear Properties of Laser Cladding Ni-Based Coatings on 6063Al. *J. Alloys Compd.* **2017**, *727*, 278–285. [CrossRef]

90. Tey, C.F.; Tan, X.; Sing, S.L.; Yeong, W.Y. Additive Manufacturing of Multiple Materials by Selective Laser Melting: Ti-Alloy to Stainless Steel via a Cu-Alloy Interlayer. *Addit. Manuf.* **2020**, *31*, 100970. [CrossRef]
91. Zhang, Y.; Wei, Z.; Shi, L.; Xi, M. Characterization of Laser Powder Deposited Ti–TiC Composites and Functional Gradient Materials. *J. Mater. Process. Technol.* **2008**, *206*, 438–444. [CrossRef]
92. Mahamood, R.M.; Akinlabi, E.T. Laser Metal Deposition of Functionally Graded Ti6Al4V/TiC. *Mater. Des.* **2015**, *84*, 402–410. [CrossRef]
93. Cao, S.; Liang, J.; Wang, L.; Zhou, J. Effects of NiCr Intermediate Layer on Microstructure and Tribological Property of Laser Cladding Cr_3C_2 Reinforced Ni60A-Ag Composite Coating on Copper Alloy. *Opt. Laser Technol.* **2021**, *142*, 106963. [CrossRef]
94. Cai, Y.; Sun, D.; Cui, Y.; Manladan, S.M.; Wang, T.; Shan, M.; Han, J. Effect of CoCrFeMnNi Transition Cladding Layer on Crack Resistance of CoCrFeMnNi + x(TiC) Composite Cladding Layer. *Mater. Lett.* **2021**, *304*, 130700. [CrossRef]
95. Thawari, N.; Gullipalli, C.; Katiyar, J.K.; Gupta, T.V.K. Influence of Buffer Layer on Surface and Tribomechanical Properties of Laser Cladded Stellite 6. *Mater. Sci. Eng. B* **2021**, *263*, 114799. [CrossRef]
96. Wang, Q.; Li, Q.; Zhang, L.; Chen, D.X.; Jin, H.; Li, J.D.; Zhang, J.W.; Ban, C.Y. Microstructure and Properties of Ni-WC Gradient Composite Coating Prepared by Laser Cladding. *Ceram. Int.* **2022**, *48*, 7905–7917. [CrossRef]
97. Li, L.; Wang, J.; Lin, P.; Liu, H. Microstructure and Mechanical Properties of Functionally Graded TiCp/Ti6Al4V Composite Fabricated by Laser Melting Deposition. *Ceram. Int.* **2017**, *43*, 16638–16651. [CrossRef]
98. Lee, Y.; Nordin, M.; Babu, S.S.; Farson, D.F. Effect of Fluid Convection on Dendrite Arm Spacing in Laser Deposition. *Metall. Mater. Trans. B* **2014**, *45*, 1520–1529. [CrossRef]
99. Fu, F.; Zhang, Y.; Chang, G.; Dai, J. Analysis on the Physical Mechanism of Laser Cladding Crack and Its Influence Factors. *Optik* **2016**, *127*, 200–202. [CrossRef]
100. El-Labban, H.F.; Mahmoud, E.R.I.; Algahtani, A. Microstructure, Wear, and Corrosion Characteristics of TiC-Laser Surface Cladding on Low-Carbon Steel. *Metall. Mater. Trans. B* **2016**, *47*, 974–982. [CrossRef]
101. Valsecchi, B.; Previtali, B.; Vedani, M.; Vimercati, G. Fiber Laser Cladding with High Content of WC-Co Based Powder. *Int. J. Mater.* **2010**, *3*, 1127–1130. [CrossRef]
102. Song, M.; Wu, L.; Liu, J.; Hu, Y. Effects of Laser Cladding on Crack Resistance Improvement for Aluminum Alloy Used in Aircraft Skin. *Opt. Laser Technol.* **2021**, *133*, 106531. [CrossRef]
103. Xiong, Z.; Chen, G.; Zeng, X. Effects of Process Variables on Interfacial Quality of Laser Cladding on Aeroengine Blade Material GH4133. *J. Mater. Process. Technol.* **2009**, *209*, 930–936. [CrossRef]
104. Baldridge, T.; Poling, G.; Foroozmehr, E.; Kovacevic, R.; Metz, T.; Kadekar, V.; Gupta, M.C. Laser Cladding of Inconel 690 on Inconel 600 Superalloy for Corrosion Protection in Nuclear Applications. *Opt. Lasers Eng.* **2013**, *51*, 180–184. [CrossRef]
105. Yu, J.; Sun, W.; Huang, H.; Wang, W.; Wang, Y.; Hu, Y. Crack Sensitivity Control of Nickel-Based Laser Coating Based on Genetic Algorithm and Neural Network. *Coatings* **2019**, *9*, 728. [CrossRef]
106. Hu, G.; Yang, Y.; Qi, K.; Lu, X.; Li, J. Investigation of the Microstructure and Properties of NiCrBSi Coating Obtained by Laser Cladding with Different Process Parameters. *Trans. Indian Inst. Met.* **2020**, *73*, 2623–2634. [CrossRef]
107. Majumdar, J.D.; Rittinghaus, S.K.; Wissenbach, K.; Höche, D.; Blawert, C.; Weisheit, A. Direct Laser Cladding of the Silicide Dispersed Titanium Aluminide (Ti45Al5Nb0.5Si) Composites. *Opt. Laser Technol.* **2018**, *106*, 182–190. [CrossRef]
108. Desale, G.R.; Paul, C.P.; Gandhi, B.K.; Jain, S.C. Erosion Wear Behavior of Laser Clad Surfaces of Low Carbon Austenitic Steel. *Wear* **2009**, *266*, 975–987. [CrossRef]
109. Saboori, A.; Gallo, D.; Biamino, S.; Fino, P.; Lombardi, M. An Overview of Additive Manufacturing of Titanium Components by Directed Energy Deposition: Microstructure and Mechanical Properties. *Appl. Sci.* **2017**, *7*, 883. [CrossRef]
110. Beuth, J.; Klingbeil, N. The Role of Process Variables in Laser-Based Direct Metal Solid Freeform Fabrication. *JOM* **2001**, *53*, 36–39. [CrossRef]
111. Javid, Y. Multi-Response Optimization in Laser Cladding Process of WC Powder on Inconel 718. *CIRP J. Manuf. Sci. Technol.* **2020**, *31*, 406–417. [CrossRef]
112. Khorram, A.; Davoodi Jamaloei, A.; Sepehrnia, R. Analysis of Solidification Crack Behavior for Amdry 997 Coating on Inconel 713 LC Superalloy by Laser Cladding Process. *Optik* **2022**, *264*, 169407. [CrossRef]
113. Shi, B.; Mu, X.; Zhan, H.; Deng, L.; Li, T.; Zhang, H. Crack Behavior of Ni60A Coating Prepared by Laser Cladding on a Tilted Substrate. *Coatings* **2022**, *12*, 966. [CrossRef]
114. Yamaguchi, T.; Hagino, H. Effects of the Ambient Oxygen Concentration on WC-12Co Cermet Coatings Fabricated by Laser Cladding. *Opt. Laser Technol.* **2021**, *139*, 106922. [CrossRef]
115. Ding, C.; Cui, X.; Jiao, J.; Zhu, P. Effects of Substrate Preheating Temperatures on the Microstructure, Properties, and Residual Stress of 12CrNi2 Prepared by Laser Cladding Deposition Technique. *Materials* **2018**, *11*, 2401. [CrossRef]
116. Bidron, G.; Doghri, A.; Malot, T.; Fournier-Dit-Chabert, F.; Thomas, M.; Peyre, P. Reduction of the Hot Cracking Sensitivity of CM-247LC Superalloy Processed by Laser Cladding Using Induction Preheating. *J. Mater. Process. Technol.* **2020**, *277*, 116461. [CrossRef]
117. Zhou, S.; Huang, Y.; Zeng, X.; Hu, Q. Microstructure Characteristics of Ni-Based WC Composite Coatings by Laser Induction Hybrid Rapid Cladding. *Mater. Sci. Eng. A* **2008**, *480*, 564–572. [CrossRef]
118. Li, G.J.; Li, J.; Luo, X. Effects of Post-Heat Treatment on Microstructure and Properties of Laser Cladded Composite Coatings on Titanium Alloy Substrate. *Opt. Laser Technol.* **2015**, *65*, 66–75. [CrossRef]

119. Chen, J.-Y.; Conlon, K.; Xue, L.; Rogge, R. Experimental Study of Residual Stresses in Laser Clad AISI P20 Tool Steel on Pre-Hardened Wrought P20 Substrate. *Mater. Sci. Eng. A* **2010**, *527*, 7265–7273. [CrossRef]
120. Chen, J.; Wang, S.-H.; Xue, L. On the Development of Microstructures and Residual Stresses during Laser Cladding and Post-Heat Treatments. *J. Mater. Sci.* **2012**, *47*, 779–792. [CrossRef]
121. Telasang, G.; Dutta Majumdar, J.; Wasekar, N.; Padmanabham, G.; Manna, I. Microstructure and Mechanical Properties of Laser Clad and Post-Cladding Tempered AISI H13 Tool Steel. *Metall. Mater. Trans. A* **2015**, *46*, 2309–2321. [CrossRef]
122. Wang, H.; Cheng, Y.; Yang, J.; Wang, Q. Influence of Laser Remelting on Organization, Mechanical Properties and Corrosion Resistance of Fe-Based Amorphous Composite Coating. *Surf. Coat. Technol.* **2021**, *414*, 127081. [CrossRef]
123. Lu, Y.; Huang, G.; Wang, Y.; Li, H.; Qin, Z.; Lu, X. Crack-Free Fe-Based Amorphous Coating Synthesized by Laser Cladding. *Mater. Lett.* **2018**, *210*, 46–50. [CrossRef]
124. Zhai, L.L.; Zhang, J.W.; Ban, C.Y. Influence on the Microstructure of Laser Cladding NiCrBSi Coatings with Electromagnetic Compound Field. *SSP* **2019**, *295*, 15–20. [CrossRef]
125. Xie, S.; Li, R.; Yuan, T.; Chen, C.; Zhou, K.; Song, B.; Shi, Y. Laser Cladding Assisted by Friction Stir Processing for Preparation of Deformed Crack-Free Ni-Cr-Fe Coating with Nanostructure. *Opt. Laser Technol.* **2018**, *99*, 374–381. [CrossRef]
126. Li, M.; Han, B.; Wang, Y.; Song, L.; Guo, L. Investigation on Laser Cladding High-Hardness Nano-Ceramic Coating Assisted by Ultrasonic Vibration Processing. *Optik* **2016**, *127*, 4596–4600. [CrossRef]
127. Qi, K.; Yang, Y.; Sun, R.; Hu, G.; Lu, X.; Li, J.; Liang, W.; Jin, K.; Xiong, L. Effect of Magnetic Field on Crack Control of Co-Based Alloy Laser Cladding. *Opt. Laser Technol.* **2021**, *141*, 107129. [CrossRef]
128. Han, X.; Li, C.; Yang, Y.; Gao, X.; Gao, H. Experimental Research on the Influence of Ultrasonic Vibrations on the Laser Cladding Process of a Disc Laser. *Surf. Coat. Technol.* **2021**, *406*, 126750. [CrossRef]
129. Wang, Q.; Chen, F.Q.; Li, Q.; Zhang, L.; Jin, H.; Zhang, J.W. Microstructure and Properties of Ni60 Alloy Coating Prepared by Electromagnetic Compound Field Assisted Laser Cladding. *Mater. Chem. Phys.* **2022**, *291*, 126678. [CrossRef]
130. Hu, G.; Yang, Y.; Sun, R.; Qi, K.; Lu, X.; Li, J. Microstructure and Properties of Laser Cladding NiCrBSi Coating Assisted by Electromagnetic-Ultrasonic Compound Field. *Surf. Coat. Technol.* **2020**, *404*, 126469. [CrossRef]
131. Wang, Q.; Zhai, L.L.; Zhang, L.; Zhang, J.W.; Ban, C.Y. Effect of Steady Magnetic Field on Microstructure and Properties of Laser Cladding Ni-Based Alloy Coating. *J. Mater. Res. Technol.* **2022**, *17*, 2145–2157. [CrossRef]
132. Zhang, M.; Zhao, G.L.; Wang, X.H.; Liu, S.S.; Ying, W.L. Microstructure Evolution and Properties of In-Situ Ceramic Particles Reinforced Fe-Based Composite Coating Produced by Ultrasonic Vibration Assisted Laser Cladding Processing. *Surf. Coat. Technol.* **2020**, *403*, 126445. [CrossRef]
133. Niu, P.; Li, R.; Yuan, T.; Wang, M.; Xie, S.; Wu, H. Laser Solid Forming Assisted by Friction Stir Processing for Preparation of Ni–16Cr–8Fe Alloys: Crack Repairing and Grain Refinement. *J. Mater. Res.* **2018**, *33*, 3521–3529. [CrossRef]
134. Aliabadi, M.; Khodabakhshi, F.; Soltani, R.; Gerlich, A.P. Modification of Flame-Sprayed NiCrBSi Alloy Wear-Resistant Coating by Friction Stir Processing and Furnace Re-Melting Treatments. *Surf. Coat. Technol.* **2023**, *455*, 129236. [CrossRef]
135. Zhang, T.; Zhou, J.; Wang, J.; Meng, X.; Li, P.; Huang, S.; Zhu, H. Effect of Hybrid Ultrasonic-Electromagnetic Field on Cracks and Microstructure of Inconel 718/60%WC Composites Coating Fabricated by Laser Cladding. *Ceram. Int.* **2022**, *48*, 33901–33913. [CrossRef]
136. Wang, L.; Yao, J.; Hu, Y.; Zhang, Q.; Sun, Z.; Liu, R. Influence of Electric-Magnetic Compound Field on the WC Particles Distribution in Laser Melt Injection. *Surf. Coat. Technol.* **2017**, *315*, 32–43. [CrossRef]
137. Kumar, A.; Roy, S. Effect of Three-Dimensional Melt Pool Convection on Process Characteristics during Laser Cladding. *Comput. Mater. Sci.* **2009**, *46*, 495–506. [CrossRef]
138. Chew, Y.; Pang, J.H.L.; Bi, G.; Song, B. Thermo-Mechanical Model for Simulating Laser Cladding Induced Residual Stresses with Single and Multiple Clad Beads. *J. Mater. Process. Technol.* **2015**, *224*, 89–101. [CrossRef]
139. Aucott, L.; Dong, H.; Mirihanage, W.; Atwood, R.; Kidess, A.; Gao, S.; Wen, S.; Marsden, J.; Feng, S.; Tong, M.; et al. Revealing Internal Flow Behaviour in Arc Welding and Additive Manufacturing of Metals. *Nat. Commun.* **2018**, *9*, 5414. [CrossRef]
140. Yao, Z.; Chen, J.; Qian, H.; Nie, Y.; Zhang, Q.; Yao, J. Microstructure and Tensile Property of Laser Cladding Assisted with Multidimensional High-Frequency Vibration. *Materials* **2022**, *15*, 4295. [CrossRef]
141. Tian, J.; Xu, P.; Liu, Q. Effects of Stress-Induced Solid Phase Transformations on Residual Stress in Laser Cladding a Fe-Mn-Si-Cr-Ni Alloy Coating. *Mater. Des.* **2020**, *193*, 108824. [CrossRef]
142. Li, C.; Xu, Y.; Jia, T.; Zhao, J.; Han, X. Numerical Simulation Research on Multifield Coupling Evolution Mechanism of IN625 Laser Cladding on Nodular Cast Iron. *Int. J. Adv. Manuf. Technol.* **2022**, *119*, 5647–5669. [CrossRef]
143. Qiao, X.; Xia, T.; Chen, P. Numerical Research on Effect of Overlap Ratio on Thermal-Stress Behaviors of the High-Speed Laser Cladding Coating*. *Chin. Phys. B* **2021**, *30*, 018104. [CrossRef]
144. Li, R.; Qiu, Y.; Zheng, Q.; Liu, B.; Chen, S.; Tian, Y. Finite Element Simulation of Temperature and Stress Field for Laser Cladded Nickel-Based Amorphous Composite Coatings. *Coatings* **2018**, *8*, 336. [CrossRef]
145. Krzyzanowski, M.; Bajda, S.; Liu, Y.; Triantaphyllou, A.; Mark Rainforth, W.; Glendenning, M. 3D Analysis of Thermal and Stress Evolution during Laser Cladding of Bioactive Glass Coatings. *J. Mech. Behav. Biomed. Mater.* **2016**, *59*, 404–417. [CrossRef]
146. Li, K.; Li, T.; Ma, M.; Wang, D.; Deng, W.; Lu, H. Laser Cladding State Recognition and Crack Defect Diagnosis by Acoustic Emission Signal and Neural Network. *Opt. Laser Technol.* **2021**, *142*, 107161. [CrossRef]

147. Cen, L.; Du, W.; Gong, M.; Lu, Y.; Zhang, C.; Gao, M. Effect of High-Frequency Beam Oscillation on Microstructures and Cracks in Laser Cladding of Al-Cu-Mg Alloys. *Surf. Coat. Technol.* **2022**, *447*, 128852. [CrossRef]
148. Yang, H.; Tang, Z.; Wan, L.; Wei, Q.; Wu, J.; Wang, A.; Jin, X.; Li, X.; Wu, Y.; Lu, G.; et al. Achieving Crack-Free CuCrZr/AlSi7Mg Interface by Infrared-Blue Hybrid Laser Cladding with Low Power Infrared Laser. *J. Alloys Compd.* **2023**, *931*, 167572. [CrossRef]
149. Zhang, P.; Chu, J.; Qu, G.; Wang, D. Numerical Simulation of Convex Shape Beam Spot on Stress Field of Plasma-Sprayed MCrAlY Coating during Laser Cladding Process. *Int. J. Adv. Manuf. Technol.* **2022**, *118*, 207–217. [CrossRef]
150. Feng, Y.; Pang, X.; Feng, Y.; Li, Z. Ultrahigh Hardness Coating with Excellent Crack Resistance Achieved by Ultrafine Eutectic. *Metall. Mater. Trans. A* **2022**, *53*, 2325–2330. [CrossRef]
151. Wang, Z.; Tan, M.; Wang, J.; Zeng, J.; Zhao, F.; Xiao, X.; Xu, S.; Liu, B.; Gong, L.; Sui, Q.; et al. Core-Shell Structural Iron Based Metal Matrix Composite Powder for Laser Cladding. *J. Alloys Compd.* **2021**, *878*, 160127. [CrossRef]

Disclaimer/Publisher's Note: The statements, opinions and data contained in all publications are solely those of the individual author(s) and contributor(s) and not of MDPI and/or the editor(s). MDPI and/or the editor(s) disclaim responsibility for any injury to people or property resulting from any ideas, methods, instructions or products referred to in the content.

MDPI AG
Grosspeteranlage 5
4052 Basel
Switzerland
Tel.: +41 61 683 77 34

Coatings Editorial Office
E-mail: coatings@mdpi.com
www.mdpi.com/journal/coatings

Disclaimer/Publisher's Note: The title and front matter of this reprint are at the discretion of the Guest Editors. The publisher is not responsible for their content or any associated concerns. The statements, opinions and data contained in all individual articles are solely those of the individual Editors and contributors and not of MDPI. MDPI disclaims responsibility for any injury to people or property resulting from any ideas, methods, instructions or products referred to in the content.

www.ingramcontent.com/pod-product-compliance
Lightning Source LLC
Chambersburg PA
CBHW062042220926
51850CB00010B/2638
* 9 7 8 3 7 2 5 8 6 7 5 8 5 *